AGROFORESTRY IN INDIA

AGROFORESTRY IN INDIA

K.G. TEJWANI

OXFORD & IBH PUBLISHING CO. PVT. LTD.
New Delhi Bombay Calcutta

Swedish International Development Authority (SIDA) has provided financial assistance for the production of this book as a low cost edition.

ISBN 81-204-0920-5

Published by Mohan Primlani for Oxford & IBH Publishing Co. Pvt. Ltd., 66 Janpath, New Delhi 110 001. Typeset by Laserpoint Technologies, New Delhi. Processed and printed at Baba Barkha Nath Printers, 26/7 Najafgarh Road Industrial Area, New Delhi 110015.

1-E4-20

Dedicated to
Mrs. Vishni K. Tejwani
for her support and
her commitment
to the cause

FOREWORD

One of my first trips abroad after joining the International Council for Research in Agroforestry (ICRAF) as its Director in 1981 was quite appropriately, to India. Some colleagues and I were invited by the Director-General of the Indian Council for Agricultural Research (ICAR) to attend and contribute ideas to a planning meeting in New Delhi of the then quite recently set up All India Coordinated Research Project on Agroforestry (AICRP-AF). It was certainly not the first official contact between ICRAF and India. My predecessor had participated in the famous ICAR meeting in 1979 in Imphal, when the AICRP-AF was conceived, and two years earlier Indian scientists had been active in the process leading to the establishment of ICRAF. And when I joined ICRAF in 1981, the chairman of its Board of Trustees, the person I signed my contract with, was the well known Indian scientist and plant breeder Dr. M.S. Swaminathan.

However, the ICAR meeting in New Delhi in the early 1980s was my first personal contact with agroforestry science and practice in India, and one that started a long and more formal relationship between ICRAF and ICAR and its various institutes involved with the AICRP-AF. In the course of this collaboration I came to visit India several times and learn to know many of the key actors in the field of agroforestry in that huge country, not only scientists and science leaders with ICAR, but also people in Government, private business and non-Governmental organizations. I also had the privilege to visit a number of ICAR and other institutes where agroforestry research was conducted in the Himalayan foothills, the Gangetic Plains, the Deccan Plateau and, may be most interestingly, the North Eastern Hill Region (Meghalaya) with its rainsoaked slopes.

The people I came to know and the places I visited in the course of my ten-year association with India, obviously only represent a small fraction of the "agroforestry reality" in the country. Still, this experience has imprinted in me the enormity of India in general and of what goes on in agroforestry science and practice in particular. This includes the variety and extent of traditional agroforestry practices, the volume of research and development work, the numbers of scientists and research institutions involved, the innumerable agroforestry development and extension projects run by both the federal and state government agencies as well as by non-governmental organizations, etc. It would obviously be an enormous, virtually impossible, task to try to summarize all what is known about, and what goes on, in agroforestry in India.

Yet, that is what the author of the present book, Dr. K.G. Tejwani, has attempted. An almost overwhelming amount of information has been compiled, literature from as far back as the 1860s until the early 1990s is lavishly quoted, and the book contains, what at least to me appears to be, a very comprehensive account of the Indian agroforestry practices and research results. A publication of this type should not be read from cover to cover as a light bedtime reading, but rather be used as a reference book or encyclopedia. It is my guess that it will be (may be it even ought to be?) the last time someone tries to produce the "ultimate", all-encompassing book on agroforestry in India—the amount of new and relevant information being produced and published every year in India is growing almost exponentially and it is virtually impossible for any individual to keep track of it. Future state-of-the-art type of publications on agroforestry in India will more likely refer to individual States, to ecoregions or to type of agroforestry practice or system.

It should come as no surprise to people familiar with agroforestry in India that the person who tried this gargantuan task is K.G. Tejwani. He has worked with land-use issues in India for more than 35 years, as a scientist, research manager (he was the Director of the Central Soil and Water Conservation Research and Training Institute in Dehra Dun when I first met him in 1980), teacher, consultant and author. He played an important role in the initial phase of the AICRP-AF and, from the early 1980's, when he retired from ICAR, and until today he has contributed substantially to the international agroforestry and land-use knowledge as a consultant and visiting scientist (to ICIMOD). Also at ICRAF we benefited greatly from his input when, in the mid-1980s we inventoried the agroforestry systems of the tropical world. I know that Dr. Tejwani has put an enormous effort into this book and that he had started already in the late 1980s with its compilation.

It is a great pleasure for me to congratulate Dr. Tejwani for successfully having brought this huge and very useful work to its conclusion.

Stockholm
30 March, 1994

BJÖRN LUNDGREN
Director,
International Foundation for Science

PREFACE

Although my association with agroforestry research dates back to 1957, the inspiration and impetus for writing this book was provided by the International Council for Research in Agroforestry (ICRAF), when I was designated as its coordinator for South Asia component of the Tropical Agroforestry Systems Inventory in 1984.

As is now well known, agroforestry practices are traditional, very old, and surprisingly, very specific to the local social, economic and agroecological conditions. The farmers, graziers, and forest dwellers have an intimate knowledge of these traditional practices—many of which, by and large, till not so long ago were socially acceptable, economically viable and biophysically sustainable. The practitioners knew this, but it was the agricultural and forest scientists and the development agencies who neither knew nor recognised the vital role that these practices played in the lives of these people.

With the establishment of ICRAF in 1978, agroforestry systems and practices started receiving greater attention. India also moved fast in this vital area of land use, since Dr. M.S. Swaminathan, the then Director-General, Indian Council of Agricultural Research (ICAR) and Secretary to the Government of India, Department of Agricultural Research and Education, was on the Board of Trustees of the ICRAF. In 1979 itself, he designated me to participate in the first workshop organised by ICRAF at Nairobi, Kenya, and he also arranged a National Seminar on Agroforestry at Imphal, Manipur, to take stock of the Agroforestry Research in India.

This was a very high-level seminar, which recommended among other things, that an All India Coordinated Research Project on Agroforestry may be formulated for implementation. I was nominated to be the Convenor of the group to formulate the project, which was finally started in 1983 and is continuing still.

It was during the national seminar, and subsequently during the process of formulating the All India Coordinated Research Project on Agroforestry that I became aware of the gaps in our knowledge about the traditional practices, of the lack of a coherent and comprehensive knowledge resource about the traditional practices, about past research, and about the future research needs and their directions. This awareness sowed "the seed for the search for knowledge" in my mind, some time towards the end of 1981. Another opportunity that came my way was in 1984, when I was invited by the East-West Center, Honolulu, Hawaii as a Fellow to work on the agroforestry practices in the Asia-Pacific region, and to formulate a proposal for a research project.

These opportunities provided the institutional base to nurture the seed to a plant for over a period of almost a decade, the outcome of which, to a very large extent, is this book. This book has been written keeping in view researchers, development agents, technocrats, non-government organisations, and students. The book should be of interest to those who are concerned with land use in agriculture, forest and soil conservation departments, agricultural and forestry colleges and universities, and international agencies.

After the historical perspective in the introductory chapter, the rest of the seven chapters cover: agrosilviculture, plantation crops under shade of trees, agricultural crops with commercial trees, shifting cultivation and other forest-based silvoagriculture practices, pastoralsiviculture, silvopastoral, and agrosilvopastoral practices. Also included in the text are detailed appendices, and a large number of references.

25/31 Old Rajinder Nagar
New Delhi 110 060, India
May 1994

K.G. TEJWANI

ACKNOWLEDGEMENTS

While the search for knowledge has to be a quest by an individual, the inspiration, ideas and support come from many individuals and organisations. For this, there are many people to whom I am indebted.

For the opportunities they provided, my grateful thanks are due to Dr. M.S. Swaminathan, former Director-General, ICAR; Dr. Kenneth F. King, former Director-General, ICRAF; Dr. Björn Lundgren, former Director-General, ICRAF; Mr. Napoleon T. Vergara and Dr. Larry Hamilton, former Senior Fellows at East-West Center, Honolulu; Dr. Collin Rosser, Former Director, International Centre for Integrated Mountain Development, (ICIMOD), Kathmandu; Dr. M.K. Muthoo, Director, Forestry Operations, FAO, Rome; Dr. Jean-Paul Lanly, Director, Forest Resources Division, FAO, Rome; and Dr. Charles J. Lankester, currently Director, Sustainable Development Network, UNDP, New York.

For the benefit of technical discussions and fruitful exchange of ideas, I want to thank, among many others, Dr. P.K.R. Nair, currently Professor of Agroforestry, University of Florida; (Late) Dr. Y.S. Rao, former Sr. Programme Advisor, Forestry Research Support Programme for Asia and the Pacific (FORSPA), FAO, Bangkok; Dr. R.V. Singh, former Director General, Indian Council of Forestry Research and Education, Dehradun; Dr. S. Chinnamani, Assistant Director General (AF), ICAR; and Professor Henry Gholz, University of Florida.

The Swedish International Development Agency (SIDA) provided a generous grant, and the ISO-Swedforest Consultants took upon themselves the responsibility of having this book published. I am especially thankful to Mr. Per Björkman, First Secretary, SIDA, Ms. Disa Sjöblom, Second Secretary, SIDA, Mr. Ulf Öhman and Mr. Mats Blakstad, Programme Coordinators of ISO-Swedforest, New Delhi, for their keen interest and constant support. I am grateful to Dr. R.V. Singh, who edited the manuscript very critically and helped in improving the content and quality of the text.

Last but not the least, there are people behind the scenes who provided logistic support. Many persons have helped in many ways, but I wish to specifically thank Mr. M.K. Puri and Ms. Fatima Pais of ISO-Swedforest Consultants, New Delhi, and Ms. Nani Bajaracharya of ICIMOD, Kathmandu.

K.G. TEJWANI

CONTENTS

LIST OF TABLES

LIST OF FIGURES

LIST OF PLATES

Plate 1: *Prosopis cineraria,* cluster bean and sesame mixed crop — arid zone

Plate 2: Lopped *Prosopis cineraria* in agriculture field — arid zone

Plate 3: *Acacia nilotica* and mustard (irrigated) — semi-arid zone

Plate 4: *Dalbergia sissoo* and wheat (irrigated) — semi-arid zone

Plate 5: *Ficus* species with wheat in the lower Himalayan region

Plate 6: *Leucaena leucocephala* and pearl millet (partly irrigated) — semi-arid zone

Plate 7: *Eucalyptus* species with cowpea

Plate 8: *Madhuca latifolia* and paddy — sub-humid zone

Plate 9: *Casuarina equisetifolia*, woodlot with sugar cane and paddy (fallow)

Plate 10: Silver oak and tea

Plate 11: Cacao mixed cropping with coconut

Plate 12: Poplar and sugar cane (irrigated)

Plate 13: Slash and burn of forest on steep land for shifting cultivation

Plate 14: *Eucalyptus* species and *Eulialopsis binata* grass

Plate 15: *Acacia nilotica* and natural grassland

Plate 16: Kerala home garden

CHAPTER 1

INTRODUCTION

1.1 HISTORICAL PERSPECTIVE

Agroforestry refers to systems of land use which combine growing or raising crops (and/or livestock) with woody plants. Land can be used to produce agricultural crops—agriculture ("agricultural production" is taken here in the broadest sense of the term, and therefore implies produce of annual, semi-annual, perennial crop plants for various uses and animals and their products); animals and their products—animal husbandry; and trees and their products—forestry. The products can be obtained by either growing or raising them separately or in different combinations (Figure 1).

Agroforestry systems are not new. In fact they are very old and traditional—for example, shifting cultivation, *taungya*, growing of tea and coffee in the shade of trees, inter-cropping under coconut, and home gardens. Most farmers in India combine agricultural crops with a few trees and livestock in their farming practices. The only thing new, at present, is the use of the term "agroforestry".

Apparently, man has been practising agroforestry since he learnt the art of cultivating agricultural crops and the domestication of livestock, and has never stopped using trees. Shifting cultivation in India is partly prehistoric, and partly a response to agro-ecological conditions in the region. In about 700 B.C., man changed from hunting and food-gathering to food production. The *taungya* system of forest plantation is reported to have been introduced in India by the British in the 19th century.

In the case of research also, research on the growing of tea bushes under the shade of trees is over hundred years old, having been started by the British in the 19th century. For a long time, there has been organised research on inter-cropping under coconut and on growing coffee in the shade. Experiments on silvopastoral systems (originally called fodder-fuel plantations) were started in the fifties. Several research institutes and universities have been conducting research on agroforestry—without using the term "agroforestry". For example, Central Plantation Crops Research Institute (CPCRI), Central Soil & Water Conservation Research & Training Institute (CSWCRTI), Central Arid Zone Research Institute (CAZRI), ICAR Agricultural Research Complex for North Eastern Hill Region, Central Sheep and Wool

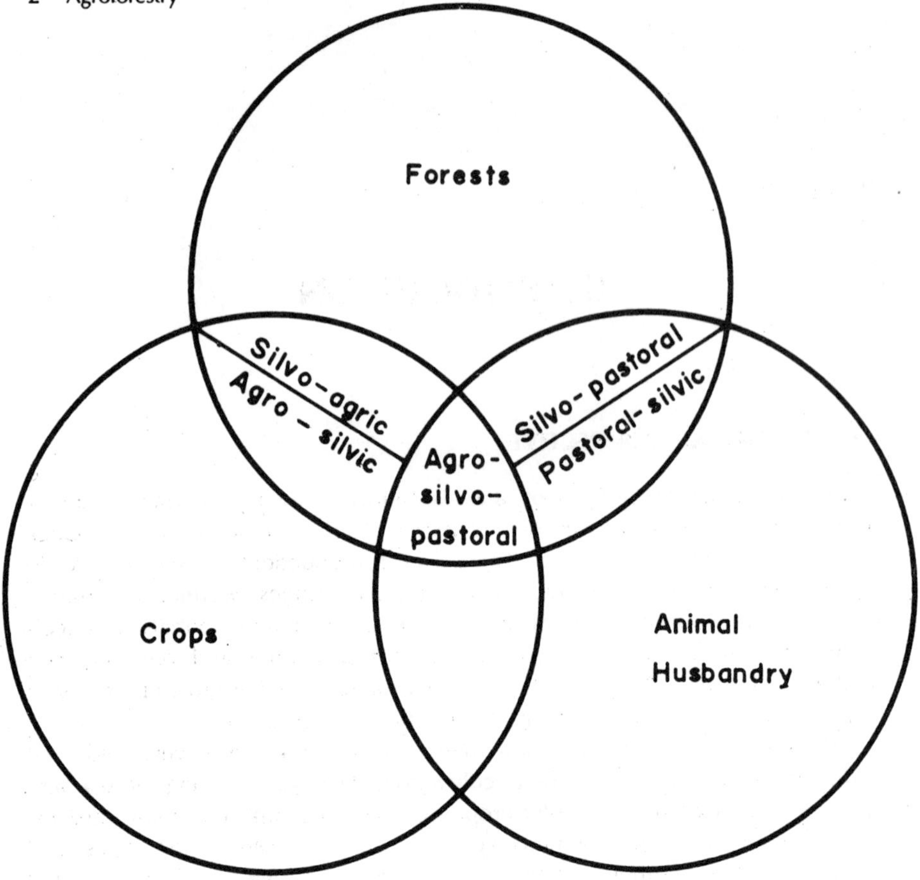

Fig. 1: Land production systems and classification of agroforestry (Adapted from Combe, 1982 by Tejwani, 1987a)

Research Institute (CSWRI), and Indian Grassland and Fodder Research Institute (IGFRI).

With the focus by FAO on forest-agriculture interface and the establishment of the International Council for Research in Agroforestry (ICRAF) in 1977, there has been an intense interest in agroforestry. In India, interest in agroforestry research has been intensified, as is evident from the series of national seminars/symposia/workshops held during the past 14 years.

The National Seminar on Agroforestry organised by the Indian Council of Agricultural Research (ICAR) in 1979, among other things recommended that "An action-oriented Coordinated Research Project on Agroforestry be drawn up by a Task Force to be constituted by the ICAR." As a follow up to this recommendation, an All India Coordinated Research Project on Agroforestry was started in 1983. This project was to operate initially at 20 centres at Research Institutes of the ICAR, and Agricultural Universities. This project was expanded considerably during the Seventh Five Year Plan (1985-1990) to cover 31 centres (Figure 2), and continues to be operational.

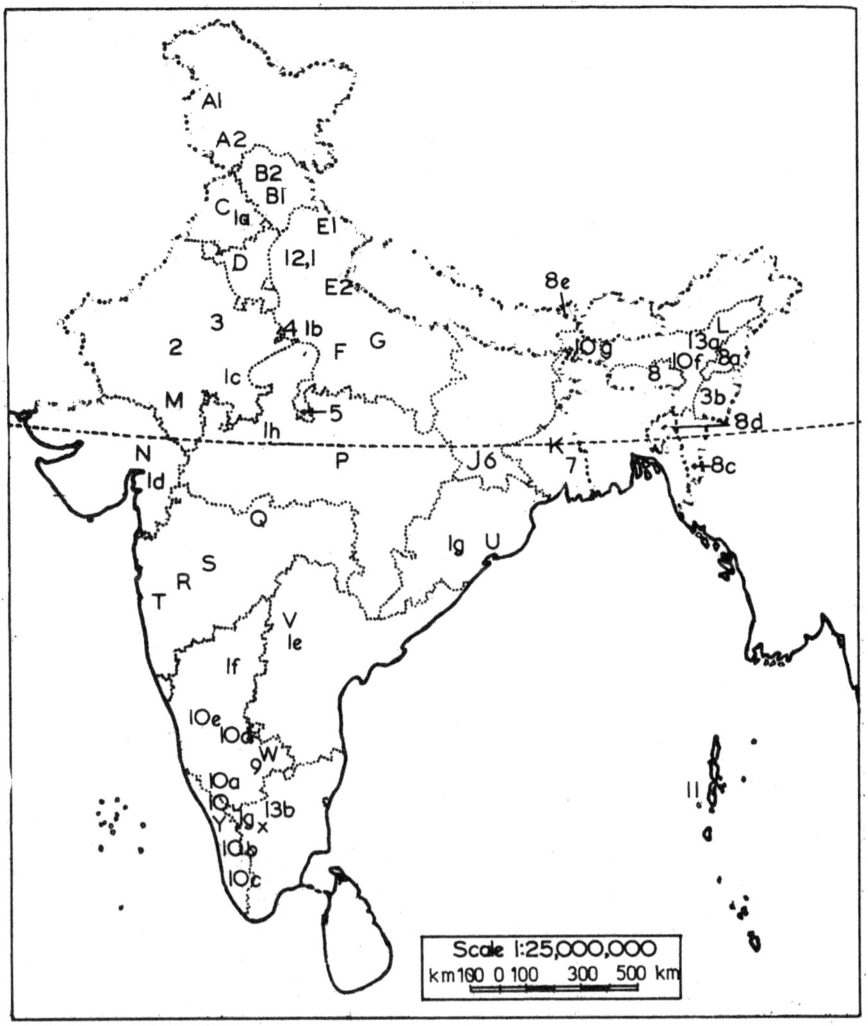

Fig. 2: Research institutes and agricultural universities where agroforestry research is carried out. For full names and locations, refer to Appendix 1 on page 173.

Research Institutions

1. CSWCRTI, and its centres at 1a to 1j; 2. CAZRI; 3. CSWRI; 4. CIRG; 5. IGFRI; 6. ILRI; 7. CIFRI; 8. ICAR Res. Complex NEH Region and its centres at 8a to 8e; 9. IIHR; 10. CPCRI, and its centres at 10a to 10g; 11. CARI for ANI; 12. FRI & C; 13a and b. TRI/TRD

Agricultural Universities

A1/A2. SK UAST; B1. HPKVV; B2. YSPUHF; C. HAU; D. PAU; E1/E2. GBPUAT; F. CSAUAT. G. NDUAT; H. RAU; J. BAU; K. BCKVV; L. AAU; M. MLSU; N. GAU; P. JNKVV; Q. PKV; R. MPKV; S. MAU; T. KKV; U. OUAT; V. APAU; W. UAS; X. TNAU; Y. KAU

1.2 AGROFORESTRY SYSTEMS

Various definitions of agroforestry have been compiled (Anonymous 1982). While conceding that in land use there are no definitions which are really comprehensive, exact and undisputed, 12 definitions have been quoted nevertheless. In summarising these definitions, it has been observed that there is a frequent mixing up of the definition, aims and potential of agroforestry. According to Lundgren (1982), a strictly scientific definition of agroforestry should stress two characteristics common to all agroforestry practices, and these should be separated from all other forms of land-use. These characteristics are:

1) The deliberate growth of woody perennials on the same unit of land along with agricultural crops and/or animals either in some form of spatial mixture and in some temporal sequence; and

2) There must be significant interaction—ecological and/or economic (positive or negative) between the woody and other components of the practice. This definition has been adopted in this book.

While promoting agroforestry, the aims, potential and positive interaction—ecological and/or economic—can be emphasized only when proved scientifically. Increased production and productivity, sustainable land management and conservation are some of the important aims of agroforestry.

Combe (1982) proposed 24 agroforestry systems (Figure 3) based on three types of association of trees (with crops, with pastures, and with both crops and pastures); two major functions of the tree component (production and protection); two spatial arrangements (regular and irregular); and two types of temporal association (temporary and permanent). This classification is still incomplete, as apiculture and pisciculture with trees belong to special classes, more difficult to classify. Interestingly, while describing various systems of agroforestry, Combe (1982) does not even mention shifting cultivation or slash and burn cultivation. Further, this classification also does not indicate whether the combination occurs simultaneously or sequentially.

In classifying agroforestry systems, Nair (1985) suggested the inclusion of agro-ecological zones and socio-economic aspects, for example, silvopastoral system for cattle production in tropical savannah or agrosilvicultural system for soil conservation in tropical savannah. Wiersum (1986) made a distinction between practices dominated by natural forests, by planted (timber) trees, by fruit or cash crop trees, and by scattered trees or trees in rows in agrosilviculture or silvopastoral systems. Vergara (1982) considered the relative allocation of land to trees, crops, pastures in various agroforestry systems. For example, in Figure 4, point A, at the extreme left shows that 100% of the land is allocated to agriculture. Similarly, on the right (point E), the area is wholly assigned to forestry. At any point between these two extremes will be found agroforestry practices with various ratios of land allocation. At any point left of point C, agriculture is more than 50%, and at any point right of point C, forestry is more than 50%.

Tejwani (1987 a) suggested a classification which among other things also took into account: (1) relative dominance of trees or crops/pastures

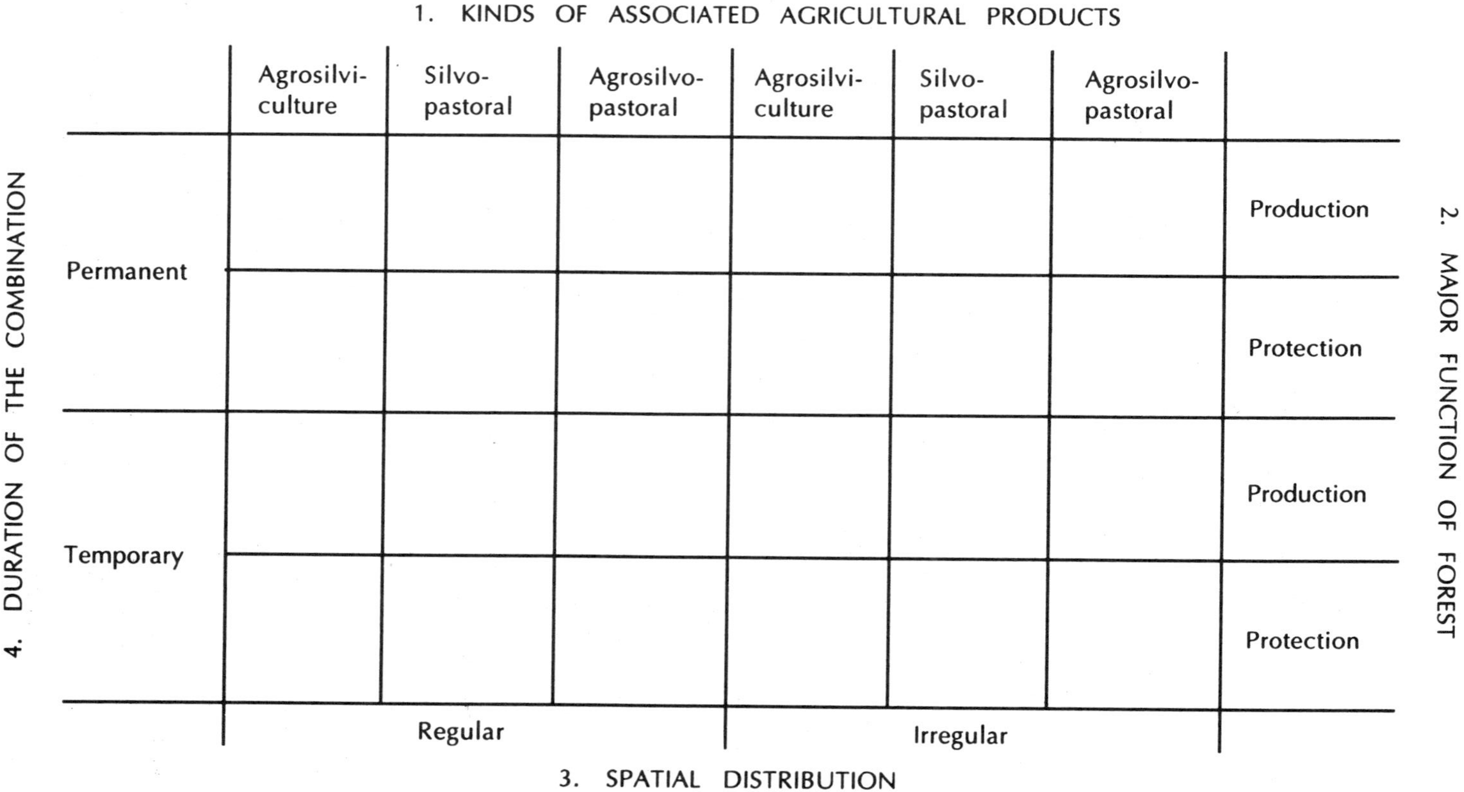

Fig. 3: Agroforestry systems classification (after Combe, 1982)

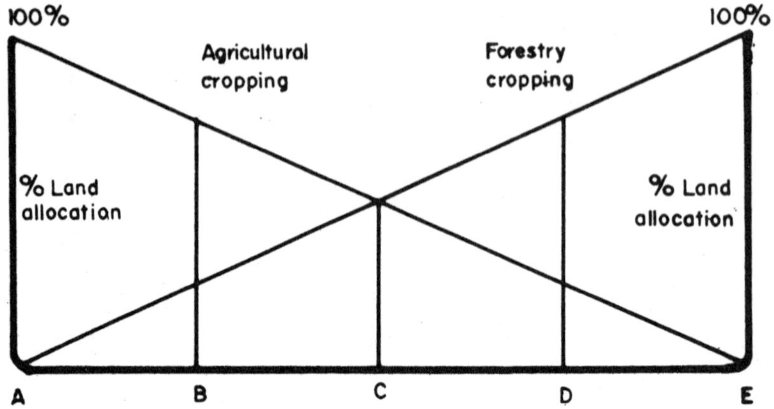

Fig. 4: Relative allocation of land under agricultural, forestry and agroforestry production systems (after Vergara, 1982)

in a practice, (2) aquaculture, and (3) non-wood forest products. For example, where trees are a major component of the land-use system, and an agricultural crop is combined with them, the system is referred to as "Silvoagriculture". Where agricultural crops are a major component of the land-use system, and trees are combined with them, the system is called "Agrosilviculture" . A combination of trees and grasses is referred to as "Silvopastoral", if the grazing land is in a forest; the system is referred to as "Pastoralsilviculture", if trees are scattered in a pasture or wasteland (Figure 1). Further, if we consider the fact that trees also grow in water (mangrove) or near water bodies (such as ponds and tanks) in which fish etc. can be grown, and keeping in view the components as defined by Combe (1982), we could have 48 types of agroforestry systems (Figure 5). This classification is comprehensive if we consider honey, flowers, food, medicinal plants, etc. harvested/collected from trees/forests, as agricultural produce.

1.3 SCOPE OF THE BOOK

For any worthwhile action-oriented or applied research, it is important to be aware, more particularly in the case of a sub-continental country like India, of the traditional agroforestry practices and research endeavours in the past so that it is easier to identify the gaps, the needs, opportunities and challenges. The cataloguing of agroforestry practices for India was first attempted in 1984 (Tejwani, 1984 b). Keeping this background in view, this book attempts to (1) catalogue the traditional practices in vogue; (2) describe the research carried out, the recent innovations, and recent interventions so far; and (3) indicate the new directions for research and development in India. While doing so, three following factors have been kept in mind in structuring the text.

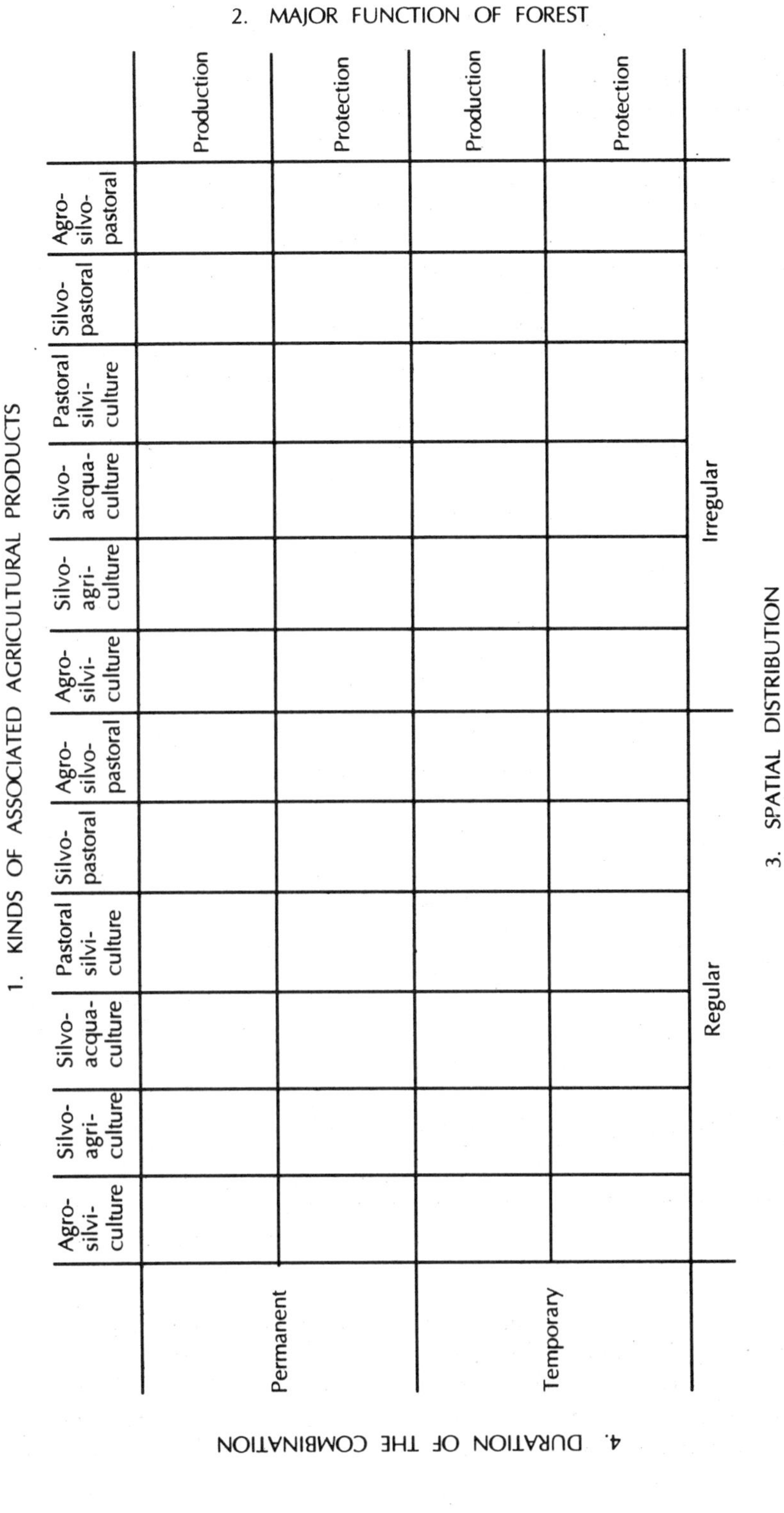

Fig. 5: Detailed classification of agroforestry systems

1) The classification of agroforestry systems suggested by Tejwani (1987 a);

2) The prevalent land-use classification in India—namely agricultural land, forest land, pasture, and wasteland; and

3) Agro-ecological situations—India is endowed with a wide variety of agro-ecological situations such as lowland humid tropics, upland sub-humid and humid tropics, arid and semi-arid tropics, temperate uplands, and alpine uplands. These have been taken into account while describing any practice.

Altogether 23 major agroforestry practices are described (Table 1).

Table 1: Structure of the agroforestry practices described

Major land classification	Dominant component	
	Crop*	Tree*
AGRICULTURE	AGROSILVICULTURE	SILVOAGRICULTURE
	(Agricultural crops with trees)	(Tree crops)
	– *Prosopis cineraria* – *Grewia optiva* and other tree spp. – *Acacia* spp. – *Eucalyptus* spp. – *Leucaena leucocephala* – Other tree spp. – Woodlots	– Coconut – Arecanut – Poplars – Horticulture
	(Commercial crops with trees)	
	– Tea – Coffee – Cacao – Pepper – Cardamom – Other crops	
FOREST	– Cardamom	– Shifting cultivation – *Taungya* – Other uses
FOREST	—	SILVOPASTORAL
WASTELANDS	PASTORALSILVICULTURE	SILVOPASTORAL
	AGROSILVOPASTORAL	

* The animal component is indicated in the practices given under various land classifications and crop/tree combinations.

Among the socio-economic aspects the density of population plays a major role in determining the types and intensity of various land uses. India has a very dense population which is mostly rural, poor, agrarian and dependent on subsistence farming. This aspect is the common thread in all the agroforestry practices described here. The local socio-economic conditions under which any agroforestry practice flourishes, or is adapted, gets highlighted at the appropriate place.

The strategy adopted in describing agroforestry practices is to first focus on agrosilviculture practices which are farm-based where common agricultural crops are grown (chapter 2); then those where commercial crops are raised (chapter 3); followed by silvoagriculture practices (chapter 4) where the tree crop is the major production component. Forest-based silvoagriculture practices are discussed in chapter 5. Pastoralsilviculture, practised on wastelands and community lands, is described in chapter 6; silvopastoral practices in forest land are described in chapter 7; these are followed by agrosilvopastoral practices in chapter 8.

CHAPTER 2

AGROSILVICULTURE PRACTICES

Agrosilviculture systems are traditionally practised in most parts of India. Two types of this system can be identified here: in one case farmers grow trees in and around fields where they grow food crops; in another case, farmers, as well as large companies grow trees in farms where the main product is commercial crops. The trees may either be grown in agricultural fields, on farm boundaries, or as separate woodlots. The extent and dimension of growing of trees by farmers in and around agricultural fields can be gauged by the fact that in Gujarat alone there were as many as 49 million trees on private lands (GOG, 1984).

2.1 TREES IN AGRICULTURAL FIELDS

Trees are grown in agricultural fields for many uses such as shade, fodder, fuelwood, fruits, and small timber. Some of the practices are very extensive and highly developed. Among them are the growing of *Prosopis cineraria* and *Zizyphus nummularia* in arid areas; *Grewia optiva* and other tree species in the Himachal Pradesh and Uttar Pradesh Himalaya, and *Eucalyptus globulus* in the southern hills in Tamil Nadu. *Acacias, Eucalyptii* and many other tree species are grown all over India.

2.1.1 *Prosopis cineraria* and *Zizyphus nummularia*

Prosopis cineraria

Distribution and Climate

The practice of growing *Prosopis cineraria* (locally known as *khejri*) in agricultural fields is extensively prevalent in the arid parts of Rajasthan and Gujarat and to some extent in the similar parts of Punjab, Haryana and western Uttar Pradesh bordering Rajasthan. The annual rainfall may vary from 200 to 500 mm. Once in 10 years the annual rainfall of the arid zone may be less than 200 mm. The practice described herein pertains to the Rajasthan desert or to western Rajasthan (Mann and Saxena, 1980; Anonymous, 1981; Shankarnarayan 1984). The mean maximum temperature during summer is 40° C. The normal maximum temperature during the southwest monsoon exceeds 35° C. After the withdrawal of the monsoon, dry and fairly warm weather prevails till the end of October, which is usually referred to as "the second summer", when the maximum temperature varies from 32-34° C; the highest diurnal variation is noticed during this season and it ranges from 15-19° C. The annual potential evapotranspiration varies from 1,616–2,063 mm (Krishnan, 1977).

Soils

Western Rajasthan desert soils have been described as "desert" soils and "grey brown" soils (Raychaudhri et al., 1963). Medium texture soils of old alluvial plains have been classified as "sierozems" in a rainfall zone of 300–500 mm and "red desert" soils in 200–400 mm rainfall zone. (Roy and Sen, 1968; Mathur et al., 1972). Soils are low in organic matter and fertility and have low water-holding capacity (Dhir, 1977). The western Rajasthan desert has been divided into 14 major land forms, each of which has a specific agricultural and water potential (Ghose et al., 1977).

Vegetation

Six vegetation types are recognised: (1) mixed xeromorphic thorn forest; (2) mixed xeromorphic woodlands; (3) mixed xeromorphic riverine thorn forests; (4) lithophytic scrub desert; (5) psammophytic scrub desert; and (6) halophytic scrub (Saxena, 1977).

According to one school of thought, *Prosopis-Salvadora* is recognised as the climatic climax of the region, while according to another there is no absolute climatic climax for the whole of the arid zone of western Rajasthan. The vegetation shows a poly-climax pattern and is adapted to the pattern of edaphic factors.

Land-use Systems

It is a revealing fact that in the arid zone as much as 45.1% of land is cultivated under rainfed conditions, (only 1.45% being double cropped), when the land basically is unsuited for agriculture. The land-capability classification of the region indicates that there are hardly any class I and II lands. Class III land occupies about 27%, class IV land occupies 16%, class V land occupies about 1.2%, and class VI land predominates and accounts for 45% of the land area (Mann et al., 1977). The surveys conducted covering 64,140 sq km (32.7% of arid Rajasthan) have shown that the present land use is inconsistent with the land capability. The capability-based recommended land use requires an increase under permanent pastures from the present 4.8-7.3% to 25-38% in the districts of Jodhpur, Jalore and Barmer.

Agriculture

More than 89% of the arid zone of Rajasthan (comprising about 60% of the Indian arid zone) will never be irrigated and will have to depend on rainfall. The crop production in the arid land is unstable and risky, leading to low and unremunerative yield levels. Pearl millet (*Pennisetum typhoides*), sorghum (*Sorghum vulgare*), sesamum, and *kharif* pulses like green gram

(*Vigna aureus*), cluster bean (*Cyamopsis tetragonoloba*) and *moth* (*Phaseolus aconitifolius*) are the main rainfed crops (Mann and Singh, 1977).

Harsh and unfavourable climatic conditions, coupled with wind-blown soils—low in organic matter and poor in moisture retention — are responsible for a sparse vegetative cover and low unstable yield of rainfed crops.

Livestock Production System

Livestock husbandry plays an important role in the economy of the arid region (Acharya et al., 1977). The livestock sector in Rajasthan accounts for 12% of the total income of the state and considering that this sector provides employment to two-thirds of the population in the arid zone, the contribution of livestock to the economy of this region would be much higher.

Cattle, sheep, goat and camel are the important livestock animals. Studies on the comparative economics of sheep and goat under range management conditions have indicated that goats are 40-160% more economical than sheep. Further, these two species do not compete with each other in utilising the naturally available resources.

The Indian Desert had 3,910,000 cattle, 750,000 buffaloes, 4,750,000 sheep, 3,490,000 goats and 360,000 camels in 1956. In 1983 the livestock population increased to 4,150,000 cattle, 1,760,000 buffaloes, 8,770,000 sheep, 7,440,000 goats and 640,000 camel (Anantha Ram and Purohit, 1992). Over the years, small animals like sheep and goat have increased at a very fast rate, while larger grazing animals like cattle, have increased at a much slower rate. The increase in buffaloes is due to the availability of water in certain parts. Moreover, buffaloes are normally stall-fed.

The productivity of livestock is generally low in the arid zone mainly due to the lack of adequate feed resources. Not only is the grazing and supplementary feed available from crop residues poor in quality, but even the total dry matter requirement is not being met. In addition, there has been a decrease in the area of pastures, barren and non-agricultural lands. Stock-watering facilities are also very inadequate, and wherever available, the water is brackish.

Forest

The Rajasthan desert has a negligible area under forest. The arid zone of Rajasthan has about 2,30,000 ha of tropical thorn forest, in the rainfall zone of 250–700 mm. This type is not only met with on flat alluvial or aeolian soils but also extends over low hill country, ravine lands and patches of saline soil. Thorny *Mimoseae* occur throughout with a number of other species, of which *Prosopis cineraria* is perhaps the most common; *Acacias* are widespread; and *Capparis decidua* is often one of the most conspicuous shrubs.

Population

By arid zone standards, the Rajasthan desert is one of the most densely populated deserts of the world—having an average density of 48 persons per square kilometre as compared with three persons in most other desert regions of the world. Though variations in the density of population occur on account of many other reasons, the availability of water is the single most important factor determining the distribution and density of population of the arid zone of Rajasthan. The population of the Rajasthan desert increased from 3.6 million in 1901 to 10.2 million in 1970 (values for the arid zone Rajasthan are available up to this period only). The enormous acceleration of growth in population and its potential rate of growth in the future is evident from the fact that in the decade of 1961 to 1971, as many as 2.2 million persons were added (Malhotra, 1977).

Pressure on Land

The consequence of the increase in population coupled with the lack of a diversified economic base, has been a tremendous pressure on land-water-plant resources. The area available to each household has been progressively decreasing since 1951 (Malhotra, 1977). It is not only that less land is available for various land uses but to meet the increasing demands of food, marginal lands which are not really suitable for agriculture are also put to the plough. Between 1957 and 1985, the gross cropped area and net sown area in the Indian desert increased by 2.66 and 2.14 million ha, respectively, while the area under all other land uses decreased, for example, 0.97 million ha under pastures and fallows, 0.18 million ha under forest, 1.8 million ha under barren and non-agriculture use, and 0.20 million ha under other land uses (Anantha Ram and Purohit, 1992). The yield per unit area is showing a declining trend. The annual linear growth rate during 1954–70 worked out to be –0.66, –4.88, –2.17, –6.08, –1.20, and –0.18% for pearl millet, sorghum, pulses, sesamum, Bengal gram and barley, respectively (Malhotra, 1977).

The size of holdings varies from 3.27 to 24.60 ha, increasing with an increase in the aridity index (Jodha, 1977). The percentage area allocated for cultivation varies from 13.3 to 88.9 in different districts; this contrasts with an average value of 78.9% of land belonging to land capability classes IV to VII.

The Practice

Prosopis cineraria trees are grown on farm boundaries as well as in agricultural fields. The crops grown are pearl millet (*Pennisetum typhoides*) and pulses during the southwest monsoon under rainfed conditions. Wheat may be grown in winter if water is available.

In a well-planned agroforestry practice of agricultural crops and *Prosopis cineraria* the trees are lopped in a systematic manner from mid-November to mid-January every year (Mann and Saxena, 1980; Saxena, 1980). In case the area is to be sown with wheat in winter, lopping is completed by the end of October or the first week of November. The trees remain dormant in the winter months. By mid-February the trees start sprouting. The tender twigs are cut in April-June when fodder is scarce for the livestock. The crowns of trees and agricultural crops grow simultaneously during the rainy season, from July to September. The trees attain full foliage by the end of September, when the *kharif* crops also start maturing. The best examples of this practice are found in the districts of Pali, Nagaur, Sikar, Jaipur and Jhunjhunu. The trees are well maintained by good farmers. In many instances, the lopping is indiscriminate and one can see unsightly and grotesque gall formations on the trunks and bases of branches. The methods of propagation, growth and utilisation of *P. cineraria* including lopping are based on traditional knowledge acquired by farmers by trial and error, through generations of experience.

In the early stages of growth, *P. cineraria* is a slow-growing tree. Generally it takes 10 to 15 years to become a tree in 200-350 mm rainfall zone. The variations in growth arise due to the variations in soil depth and total annual rainfall. Tree growth is better in high rainfall areas (350-450 mm) than in low rainfall areas (100-200 mm) and shallow soils. The tree assumes a stunted growth in shallow soils with hard *kankar* pan. The slow rate of growth of the aerial part is one of the adaptations of this tree to withstand drought. It tries to establish itself first with its strong tap root system and then allows shoot growth. The tap root travels vertically down in the soil and ensures the establishment of the plant during the first ten years. *P. cineraria* root can penetrate up to a depth of 36 m. The very deep root system helps this plant to withstand drought successfully (Shankarnarayan et al., 1987). It also has high capacity to withstand frost (Saxena, 1980), and alkalinity (pH 9.8) (Khan, 1955).

Prosopis cineraria is well distributed throughout the plains of western Rajasthan on sandy loam, loam, and sandy clay loam soils with a hard *kankar* pan at varying depths from 50-150 cm (Shankar, 1980 b). Saxena (1980) has described the habitat, distribution and associates of *P. cineraria*. On the older alluvial plains (with 60-90 cm soil depth) there are 15-20 trees/ha. The growth attributes will be 60-90 cm girth, 10-14 m height and 50-60% dominance, in the prevalent plant community of *Prosopis cineraria-Zizyphus nummularia-Capparis decidua*. On the older alluvial plains with hummocks, *P. cineraria* is sparsely distributed (10-15 trees/ha) between the hummocks. *Tecomella undulata* and *Balanites aegyptiaca* are the main tree associates of *Prosopis*. In the younger alluvial plains *Acacia nilotica-Prosopis cineraria* is the most prevalent plant community. *Prosopis cineraria* has a height of 10-15 m, 40 cm dbh and a density of 25-45 trees/ha. Its chief associate *Acacia nilotica* has a dominance of 13-25%. The lower piedmont plains support *Salvadora oleoides-P. cineraria-Euphorbia caducifolia* community where *Prosopis cineraria* has 25-35% relative dominance. The trees have 6-8 m height, 15-40 cm dbh, and density is 35-50 trees/ha. In the buried piedmont plains, the common plant community is *P. cineraria-Acacia senegal-Capparis decidua*. *Prosopis* has 50-65% dominance, 20-35 trees/ha with 8-10 m height and 20-28 cm dbh. In the stabilised sand dunes in a rainfall zone of 250-400 mm, the density is 5-25 trees/ha. The interdunal plains have a good density (40-60 trees/ha) of *Prosopis*, while undulating hummocky areas show comparatively low density (20-40 trees/ha). Common communities of interdunal plains are *P. cineraria* alone; *P. cineraria* with *Salvadora oleoides* or *Capparis decidua* or *C. polygonoides*. In the low rainfall areas (150-250 mm) the interdunal plains have very poor *P. cineraria* growth—the trees are stunted (3-4 m in height), less branching (two to three main branches) and 3-10 trees/ha. In the high rainfall zone, the density is 50-80 trees/ha. Density of *P. cineraria* increases from the western to eastern Rajasthan. The rainfall belt between 250 and 400 mm shows higher density.

In spite of the higher percentage of cultivation on flat and undulating plains, and its continuous lopping, even removal, *Prosopis* continues to dominate the tract. The plant has a high regeneration capacity and it manages to survive the odds by its high drought-hardiness.

Moisture Relations of P. cineraria

It is roughly estimated that 50 trees of *P. cineraria* in a hectare of land may use as much as 222 mm of water annually, in an area which receives an average rainfall of 356 mm (Lahiri, 1980). Roots of this tree have been excavated up to 8-10 m but thick roots with secondary growth were found to go further down. The roots penetrate the *kankar* layer also. It has been further observed that the shallow-rooted ground vegetation, which grows only during the rainy season (July-September) under the *P. cineraria* community, consumed about 162.9 mm of water, which was about the same as the precipitation (163.8 mm) received during the growing period. It was therefore concluded that the established trees with deep roots rarely compete for moisture with shallow-rooted grass cover.

Products

The utilisation of each and every part of *P. cineraria* fulfills a variety of needs of the rural people (Purohit and Wajid Khan, 1980; Muthana, 1980). It provides fuelwood, timber, forage, food, fruit, tannin, gum and medicine. Branches, twigs and other parts after lopping are used as fuelwood. Fuelwood is of high calorific value (sapwood 5,003 kcal/kg) and is used for making charcoal. The wood ash contains 31% of soluble potassium salts and may be used as a source of potash.

Its thorny twigs and branches are used for fencing the fields. The green and dry pods are used as vegetable, and the ripe pods are considered the dry fruit of the desert. One tree may give as much as 5 kg of dry ripe pods. The pods are fed to milch cattle. The leaves of *P. cineraria* provide nutritive green fodder; the dry leaves are preserved and fed to livestock, especially goats, camel and sheep.

Lopping of trees is practised in relatively high-rainfall areas. In low-rainfall areas, the leaves are harvested by defoliating the branches with leather gloves. This process removes 75% of the leaves and does not damage the branches. The dry leaves are also sold and they fetch a good value. Apart from dry leaves (*loong*) small slender green twigs with leaves are marketed in the form of small 1/2 kg bundles. The leaves have a high nutritive value.

Prosopis cineraria trees provide leaf fodder from the tenth year onwards and may go on producing up to about 200 years. A moderately sized tree yields nearly 25-30 kg of dry leaves per annum. In the green stage, the moisture content of the leaves may be as much as 60%. The dry leaves contain 93.2% dry matter, 92.1% organic matter and 16.8% crude protein.

The farmers use traditional knowledge when lopping the trees. The experimental information on lopping of *P. cineraria* in the Rajasthan desert indicated that the intensity of lopping, that is, complete, or two-third, or one-third crown lopped, did not adversely affect the growth parameters of the tree (mean height, and girth at breast height) but reduced the yield of green-leaf fodder. Lopping after every four years gave more yield of green-leaf fodder than lopping every year (Bhimaya et al., 1964). Studies conducted in the arid region of Haryana have indicated that annual lopping of *P. cineraria* trees resulted in the maximum forage yield, without detriment to the growth of the trees (Srivastava, 1978). It was further observed that the yield of forage from *P. cineraria* trees with a girth below 45 cm was significantly lower than that from trees with a girth of 45 cm and above. It is therefore suggested that only trees with a girth of 45 cm and above may be lopped.

During the migration of livestock, shepherds cut the branches of trees and the animals relish this fodder very much.

The flowers are used as medicine for blood purification, curing skin diseases and boils, and for the prevention of miscarriage. The bark is used as medicine against summer boils, leprosy, dysentery, bronchitis, asthma, leucoderma, piles, tumours of muscles and wandering of the mind.

The bark as well as the galls formed on the branches are used for tanning.

The roots are used for making cot frames, handles for agricultural implements, butter-churning sticks and rakes, and upper parts of bullock carts, as the roots are stronger than stems and are less affected by insect pests.

Its gum is used as food in sweets given to nursing mothers.

The indigenous trees and shrubs in the arid zone of Rajasthan which have been identified as providing valuable forage on the basis of palatability ratings are:

Acacia nilotica; Prosopis cineraria; Salvadora oleoides; Zizyphus nummularia; Acacia senegal; Albizia lebbek; Anogeissus rotundifolia; A. pendula; Calligonum polygonoides; Azadirachta indica; Grewia tenax; G. spinosa; Prosopis juliflora and *Tecomella undulata* (Ganguli et al., 1964).

Dried and stored leaves of *P. cineraria* and *Zizyphus nummularia* are the mainstay of both the rural and the organised dairy sectors in the western Rajasthan, involving both cattle and goats.

The uses to which other associates of *P. cineraria* are put are:

Acacia nilotica and *A. jacquemontii* provide fuelwood and timber; *Tecomella undulata* is used for timber; *Acacia senegal* provides seeds and fuelwood; *Capparis decidua* provides fruits for vegetable; *Calligonum polygonoides* gives fuelwood, and its flowers are used as a vegetable.

Animals

Animals are a part of the practice. Trees are managed to provide fodder for the livestock; animals also graze the grasslands and receive fodder from the agricultural crops.

On the strength of the evidence reported so far, it is concluded that the goat is the most efficient, followed by the camel, and then by sheep in utilising the nutrients of *P. cineraria* leaves. It is further concluded that the leaves which are collected in summer are somewhat better utilised by the livestock than those collected during winter.

Interactions

The very fact that *P. cineraria* is there, and that its occurrence is encouraged in all the cultivated fields and village grazing grounds, shows that its usefulness is generally and widely accepted by land owners who have a strong conviction, based on their own observations, that crop growth under it is better than that in the adjoining areas under the same management conditions.

Studies conducted have shown that the soil under *P. cineraria* has more organic matter, total nitrogen, total phosphorus, total potassium, available phosphorus and potassium and available micronutrients (Zn, Mn, Cu and Fe) and slightly lower pH and electrical conductivity than the soil under field conditions without *P. cineraria* (Aggarwal, 1980). On the basis of the availability limits the soils under *P. cineraria* can be categorised as those with high available phosphorus for crop production, whereas desertic soils are generally quite low in available phosphorus.

Soil under *P. cineraria* has more population of bacteria, fungi, actinomycetes and nitrifying bacteria than that under *Albizia lebbek*, *Tecoma undulata* and *Prosopis juliflora*.

The observed improvement of plant growth under the *P. cineraria* tree is therefore due to the combined action of better soil moisture and soil fertility. The additional possibility of contributions of bird droppings, leaf/pod fall, dung and urine of shade-seeking livestock may also not be ruled out.

People are aware of the positive interaction of trees and crops in the Rajasthan desert, so much so that they have coined many "sayings" (Purohit and Wajid Khan, 1980; Shankar, 1980a). For example it is said that if the growth of pods in *P. cineraria* is profuse, the next wheat crop will be better; if *Capparis decidua* yields good fruit, then the sesamum crop will be better; if *Calotropis procera* does well, then possibilities of good cotton are more; and if *Calligonum polygonoides* blooms then there is a possibility of good times. There is another saying that when the growth of pods of *P. cineraria* is good, there will be a good crop year, and if the growth of the fruit of *Zizyphus nummularia* is more, then it will be a bad crop year.

Resource Input and Utilisation

There is no information available about the inputs with respect to the tree component. Some information is available for agricultural crops only. Fertiliser use in arid and semi-arid drylands has been almost non-existent, due mainly to uncertain rainfall and cultivation of non-responsive crop varieties. As such, the earlier belief of dryland farmers, that fertiliser application was injurious to dryland crops, was justified. With the availability of fertiliser-responsive varieties of dryland crops and the knowledge gained on the quantity, time and method of nutrient application, judicious use of fertilisers can offer enormous possibilities of increasing dryland crop production under normal and above-normal rainfall situations, and stabilising it in years of subnormal rainfall (Mann and Singh, 1977). Research carried out so far on fertiliser use on drylands has clearly indicated the beneficial effects of optimum nutrient application to most crops (pearl millet, sorghum, green gram, and sesamum).

Protective and Service Aspects

The practice has many indirect protective, social and economic advantages. For example, apart from increasing or improving soil fertility, *P. cineraria* acts as a soil binder and decreases the velocity of hot summer winds (Purohit and Wajid Khan, 1980). It provides shade to humans, livestock and birds in the scorching heat of summer. It stands out with its green foliage in acute dry periods, its soothing effect to the eyes in the dazzling sunshine of the arid zone, and its most aesthetic value from among all the trees of the region.

It provides a source of revenue during droughts, and builds up the balanced economy and self-sufficiency of the farmers. The value of a field is determined by the number of *P. cineraria* trees it has—the more the trees, the higher is its value (Purohit and Wajid Khan, 1980; Srivastava, 1984). In fact, *P. cineraria* trees are also a cash crop and are economically more stable than agricultural crops. Farmers therefore protect and preserve them with great care.

Historically, *P. cineraria*, has played a significant role in the rural economy in the northwest arid region of the Indian subcontinent. The *P. cineraria* tree is very much a part of the cultural and socio-religious life of the people (at marriage, death, religious rites, and for witchcraft). The shrines of several gods are generally situated beneath this tree. References to *P. cineraria* occur in the Rig Veda, (the oldest of the four Vedas), Atharva Veda, Ramayana, Mahabharata, and many other Indian scriptures (Shankar, 1980 a; Purohit and Wajid Khan, 1980).

Improvement of the Practice

The practice can be degraded if the trees are not properly lopped. No case history of its breakdown has been reported. The Central Arid Zone Research Institute has described and explained the practice at length. It could perhaps be improved by having fast-growing and high-yielding varieties of *P. cineraria*. Some fast-growing strains of *P. cineraria* have been observed to achieve a mean height of 3.99 m (range 2.30 to 5.20 m) in seven years (Shankarnarayan et al., 1987). Indian scientists have initiated the work to study the morphological and genetic variability in *P. cineraria.* The Central Arid Zone Research Institute is selecting elite phenotypes for clonal multiplication and related investigations (Koul et al., 1991).

It is reported from the arid zone of Africa, that *Acacia albida* has a synergic impact on crops grown in association with it. Since it is relatively fast growing, its impact on crop yields has been compared with that of *P. cineraria* (Shankarnarayan et al., 1987). It was observed that at the end of three years of growth *A. albida* achieved a height growth of 89 to 128 cm, as compared to 13 to 16 cm of *P. cineraria. Prosopis cineraria* did not adversely impact on the yield of associated crops such as green gram and cluster beans. However, the yield of cluster beans was adversely impacted during the third year by *A. albida.* This adverse impact was, among other factors, attributed to the faster growth of *A. albida,* as well as its relatively shallower, horizontally-spreading root system. Unlike the *P. cineraria* roots which go very deep, the tap root of *A. albida* after penetrating 45 cm, allowed the emergence of about a dozen laterals at approximately 25 cm depth, and these grew horizontally up to 5.5 m in length. These shallow, lateral roots utilise much more soil moisture and nutrients and thus create an adverse edaphic climate for crops.

The scope of increasing production by the scientific management of trees and crops together as a system, has yet to be explored. Some of the identified research needs are:

Selection of fast-growing trees of *P. cineraria*; evaluation of fast-growing trees of *P. cineraria* with respect to high foliage and food production commensurate with high crude protein and high digestibility, and low tannin content of leaves; improvement and development of already existing spineless trees of *P. cineraria*; package of agroforestry to be developed with respect to lopping cycle, lopping intensity, plant spacing, biological control of gall formation and insect pests, etc.; and economic evaluation of the whole practice with respect to direct and indirect benefits.

Zizyphus nummularia

Bordi (*Zizyphus nummularia*) grows abundantly and wild in the arid and semi-arid zones. The crops grown are the same as in the case of *Prosopis cineraria.* The harvesting of the *bordi* shrub coincides with the season of its leaf fall and after the *kharif* crops are harvested. In a good or normal rainfall year, the farmers get a double harvest, i.e., the crop as well as *bordi* leaves for fodder, while in drought years, when the crop fails, the cultivator gets only leaf fodder. The plants are cut to the ground level by a sharp implement and heaped together in one bundle. For one week all such bundles are put at one place for drying in the sun. Once dried, the leaves are easily separated from the twigs, by beating with a stick. These leaves or *pala* are collected and stored for feeding to cattle during lean periods (Saxena, 1981; 1984). The yearly cut shrubs in agricultural fields do not produce any fruit. the *Bordi* shrub has a remarkable power of regeneration through its root suckers. The cut stumps start coppicing by February or March. The sprouting continues through April to June, indicating that the plant growth is independent of surface-soil moisture. The spring sprouts are grazed during the summer months (April to June).

Distribution

Bordi (*Zizyphus nummularia*) is one of the most common shrubs which occupies all the habitats. Flat alluvial plains with fairly deep (90-120 cm) sandy loam soils, with *kankar* pan below, support *Prosopis cineraria-Zizyphus nummularia-Capparis decidua* as the most prevalent community. Generally there is a moderate to high shrub density (230-500 shrubs/ha) of *bordi*. Due to the yearly cutting of this shrub for leaves, it does not impart any visual impact of its density. Flat alluvial plains with clay loam to clay soils and a *kankar* pan at 45-60 cm show high shrub density (4% cover) in the semi-arid region. Shallow soils with *kankar* pan at 30-45 cm are generally dominated by *bordi* shrubs. The gravelly plains where sheet wash has exposed the *kankar* pan, are dominated by a community of *Z. nummularia* and *Capparis decidua*; there are about 120-150 *bordi* plants/ha which contribute 8% crown cover (Saxena, 1981).

In piedmont plains, and eroded rocky surface, *bordi* and *Capparis decidua* are equally dominant in the community of *C. decidua* and *Z. nummularia*. These lands mostly serve as common grazing grounds. Sandy, undulating, hummocky plains and interdunal plains have a few scattered shrubs of *bordi*. In protected forests on low hills and piedmont zone, *bordi* is an important associate of *Acacia senegal*. The shrub density varies from 80-350 plants/ha but with small crown cover (5-7%). In unprotected areas, with a community of *thor* (*Euphorbia caducifolia*), *bordi* has negligible presence (10-50 shrubs/ha) due to its severe exploitation (Saxena, 1981).

Products

All parts of *bordi* are used (Purohit and Wajid Khan, 1981). Leaf fodder is the most important product in agricultural lands. The leaves are valuable fodder for camel, goat, sheep and cattle. The leaves can be given either alone or with some *bhusa* (straw/stalk of crops—pearl millet, wheat) and with leaves of *Prosopis cineraria*. The *pala* fetches a good price according to the season, and its availability.

Thorny twigs are used for fencing, for making walls of village huts (along with branches of *Calotropis procera*), doors and gates of cattle sheds, courtyards, fields, storage yards (along with branches of *Prosopis cineraria* and *Calotropis procera*), screens (along with bamboo sticks) and for soil conservation (against wind erosion).

The bark of the roots of *bordi* is used for medicinal preparations for weak sheep due for lambing, for providing strength to humans, and as a digestive and nutritive tonic for bullocks.

Its fruit is eaten by all—human beings, wild animals and birds (Prakash, 1981).

Due to its economic importance, it plays a vital role in the social, religious and cultural life of the people of the western Rajasthan (Purohit and Wajid Khan, 1981). It finds mention in the ancient scriptures (Shankar, 1981). If *Prosopis cineraria* is the king of the desert, *Zizyphus nummularia* is the "queen of the desert".

The above description does not imply that *P. cineraria* and *Z. nummularia* practices are mutually exclusive. In fact, both these species can be present on the same field. The dominance of any one of them is determined by the total annual rainfall. In areas with less than 150 mm annual rainfall, *Z. nummularia* is dominant and a few *P. cineraria* trees are present. The number of *P. cineraria* trees increases as rainfall increases, and simultaneously the number of *Z. nummularia* bushes decreases. With *P. cineraria* as the dominant species a few trees of *Ailanthus excelsa* or *Acacia leucophloea* are also present.

2.1.2 *Grewia optiva* and Other Tree Species

Distribution and Climate

This system is practised in Jammu & Kashmir (J & K), Himachal Pradesh (HP), and Uttar Pradesh (UP) (Garhwal and Kumaon Regions) Himalaya at elevations of about 550 to 2,300 m. The climate is sub-tropical to temperate, according to the elevation, with distinct rainy, winter and summer seasons. It is sub-tropical in the Siwalik foothills and up to 1,000 m elevation in the outer Himalayan ranges. The mean annual rainfall is 1,000-2,500 mm, mostly from the southwest monsoon, during June to September. The temperature varies widely due to factors of location, altitude, topography, etc.—maximum temperature of up to 40° C at lower altitudes, and minimum below freezing at higher altitudes, is recorded.

Edaphic Conditions

The topography is hilly, undulating, sometimes rugged; higher slopes are steep to precipitous. Some of the valleys are broad and flat, for example, Kangra in Himachal Pradesh. The soils are very variable, generally shallow, sandy loam to clay loam/clay.

Land-use Systems

Major land-use systems are agriculture (including horticulture), forestry, agroforestry (agrosilviculture, silvoagriculture, silvopastoral, pastoralsilviculture, agrohorticulture, silvohorticulture, multi-purpose forest tree species, and agrosilvopastoral). Agriculture is mainly at the subsistence level with some cash crops like potato, ginger, chillies and beans.

Socio-economic Conditions

The average farm size is about 1.6 ha in Himachal Pradesh and one ha in the Chakrata hills of Uttar Pradesh. The size of agricultural fields is also small—in fact these small holdings are further fragmented and could be at many locations, and at each location they could be further fragmented

(Shah, 1981; Ghildyal, 1981). Farm lands are mostly exclusively held by individual farmers, and sometimes on a tenancy/share cropping basis.

Density of population is 77 and 95 persons per sq km in the Himachal Pradesh and Uttar Pradesh Himalaya respectively as per the 1981 census (PC, 1982; Census of India, 1982).

The Practice

The major agricultural crops grown are maize, paddy, wheat, barley and potato; off-season crops are beans, tomatoes, cabbage and ginger. In horticulture, the main fruits are apple, plum, peach, apricot and mango (at lower altitudes). The fruit trees are also grown in agricultural fields in an agrohorticulture system.

The more important trees grown by the farmers are *Grewia optiva* (300-2,500 m), *Morus serrata* (1,000-2,700 m), *Celtis australis* (500-2,500 m), among others for leaf fodder, fibre, fuel and timber. Singh (1991) reported 28 tree species growing in agricultural fields and boundaries in Sirmaur district of Himachal Pradesh. Among these, four species (*Grewia optiva, Celtis australis, Toona ciliata* and *Bauhinia variegata*) are the most common, and constitute 79.1% of the total basal area of the trees. Among these *Grewia optiva* was the most common and constituted 43.3% of the total basal area of the trees. All the trees have multiple uses and are retained for fodder, fuelwood, fruit and timber. Singh (1991) reported that 61-72% of the trees were mainly fodder trees, 13-28% were mainly timber trees and 1-15% were fruit trees. It may be noted that fuelwood is provided by all the trees. The number of trees retained per hectare varied widely (from 10 to 182) depending upon the soil fertility, irrigation facility, needs of the household, size of holding, and distance from the forest. The number of trees per household also varied widely (from 12 to 129), this being determined by the size of holding—the farmers with smaller holdings generally retaining more trees per hectare of cultivated area than those with bigger holdings (Singh, 1991).

Grewia optiva

A small- to medium-sized deciduous tree; flourishes in regions with mean annual precipitation of 1,200-2,500 mm; strong light demander, can withstand drought and frost; pollards and coppices well. Its timber is used for axe-handles, shoulder poles, cot frames etc. Its bark yields a fibre which is used for basket making; its fruit is edible. It is seldom used as fuelwood because of its unpleasant odour (Gupta, 1956). Its leaves provide very nutritious fodder; the leaves and edible green twigs are palatable, nutritious and easily digestible. The leaf fodder, when fed with straw or other inferior dry roughage can profitably substitute concentrates. Nutritious young leaves are converted into a protein-rich meal after drying in the sun. The species is raised by planting seedlings or stumps, usually at 8 m spacing, in single rows on field bunds/terrace risers. It is very heavily lopped especially during winter, when no other green fodder is available (Ram Parkash, 1983; Ram Parkash and Hocking, 1986).

Celtis australis

A moderate- to fairly large-sized deciduous tree which can grow over a wide altitudinal range, 500-2,500 m. Average annual rainfall in its habitat varies from 1,200 to 2,500 mm; precipitation at the lower altitudes is entirely in the form of rainfall from the southwest monsoon as well as in winter; whereas at higher altitudes in the temperate zone it is partly in the form of snow in winter. It is sensitive to drought and fire, fairly frost-hardy and is a moderate light-demander. It pollards and coppices well. It is raised primarily for fodder but is also used as fuelwood, and for making tool handles and ploughs. The leaves make good fodder. Its chemical composition varies with season. The summer-lopped fodder has a higher nutritive value than autumn-lopped fodder. The species is raised by seedlings or by rooted root-shoot cuttings (Ram Parkash, 1983).

Morus serrata

A moderate- to large-sized deciduous tree, found in the lower western Himalayan mixed forests at 1,200-2,700 m elevation. It is a tree of sub-tropical to temperate climate; grows in a rainfall zone of 660-2,970 mm, part of which is received as snow at temperate high elevations; it is frost-hardy, a good coppicer and produces root suckers.

Morus serrata is raised primarily for fodder and timber, but also for fuelwood. Its leaves make good fodder; they are a valuable base for sericulture. Its wood is excellent for furniture, carving and cabinet work; it is also used for toys, sport goods and agricultural implements. The fruit is sweet and edible. The tree is raised on and around farm lands by seedlings, stumps or rooted branch cuttings (Gupta, 1956; Ram Parkash, 1983; Ram Parkash and Hocking, 1986).

Farmers also raise *Bauhinia variegata, Robinia pseudacacia, Albizia stipulata, Moringa oleifera, Quercus* spp. etc. for fodder and fuelwood. The Himalayan oaks (*Quercus* spp.) are multipurpose trees yielding fodder, fuelwood, charcoal, timber, vegetable (juvenile current year leaves of *Q. leucotrichophora*), sweet acorns of *Q. ilex*, pulp of *Q. leucotrichophora, Q. floribunda, Q. semecarpifolia, Q. aliena* and *Q. serrata,* tannin of *Q. leucotrichophora* and *Q. floribunda,* medicine from *Q. leucotrichophora,* acorn oil from *Q. ilex., Q. leucotrichophora* and *Q. floribunda*, cattle and poultry feed from acorns of *Q. leucotrichphora, Q. floribunda* and *Q. ilex* after oil has been extracted; and leaves of many oaks for raising *tussar* silkworm (Ghildyal, 1991).

There is no regular management of all the above tree species. *Grewia, Celtis* and *Morus* are ready for production of leaf fodder at the age of about 10 years and continue to be lopped for about 30 years. *Grewia* trees are lopped in winter, whereas *Celtis* and *Morus* trees are lopped in summer, thus ensuring a green fodder supply throughout the year.

Animals

are part of the practice, since the trees are grown to provide fodder for the livestock. The common domestic animals are sheep, goat, buffalo, kine and some ponies. Each family on an average, keeps about four cattle. These domestic cattle are partially stall-fed with grass (green or hay), and/ or leaf fodder, with very little concentrates. For most of the time they are let out in the village wastelands, adjoining forest areas, grass reserves (after harvesting grass) and fallow agricultural fields (Ram Parkash, 1983).

Arrangement of Components

The agricultural land holding of about one hectare, is highly fragmented and usually terraced. The agricultural crops mentioned above are raised as per the seasons, rotations and needs of the farmer. Tree species mentioned above are raised on field *bunds* at a linear spacing of 5-8 m in the row. The entire length of the *bunds* and terraces is not planted systematically; hardly one-third of the area is planted. The average number of trees per farm holding is about 30. There is no specific or defined admixture between the trees and crops.

Interactions

Very little is known about the interactions between trees and crops. The magnitude of interactions and their effects is also not quantified.

It is true that trees do compete with agricultural crops for nutrients, soil moisture and light, but the planting is so arranged in respect of location, spacing, and direction that perhaps and hopefully, trees cause the minimum adverse effect on the crops. Recent studies have shown that the yield of unirrigated wheat (grown in *rabi*) and paddy (grown in *kharif*) crops decreased consistently and progressively when *Grewia optiva, Morus alba* and *Eucalyptus hybrid* were grown along with them at 4 × 5 m spacing (Table 2). Over a period of 13 years there was an average loss of 29, 29 and 28% in the yield of wheat when grown in association with *Grewia optiva, Morus alba* and *Eucalyptus hybrid* respectively (Khybri et al., 1992); the average yield of wheat grain without trees being 2,249 kg/ha. In the case of paddy, the average yield reduction (for eight crops) was 28, 34 and 30% when grown in association with *Grewia optiva, Morus alba* and *Eucalyptus hybrid,* respectively; the average yield of paddy without trees being 2,275 kg/ha. *Grewia optiva* and *Morus alba* were lopped after six and five years, respectively after planting. On an average *Grewia optiva* yielded 0.25 tonnes of air-dry leaf and 1.08 tonnes of air-dry branches per hectare per annum for eight years when it was harvested at the age of 13 years; *Morus alba* yielded 0.28 tonnes of air-dry leaves and 1.28 tonnes of air-dry branches per hectare per annum for nine years, when it was harvested at the age of 13 years.

Eucalyptus hybrid was harvested at the age of 10 years, when it gave a total harvest of 32.0 tonnes/ha.

The economics of growing agricultural crops with and without trees were worked out. The total yield of wheat for 13 years, paddy for eight years, yield of trees at harvest, and loppings for eight years of *Grewia optiva* and nine years for *Morus alba* were treated as income; cost of cultivation of crops and formation of trees were treated as costs. It was observed

Table 2: Economics of different agrosilviculture treatments (rotation of 13 years)

Treatment	Yield (kg/ha) of crops		Total Income from crops (Rs.)			Total Income from trees (Rs.)	Total income (Rs.)	Total cost of cultivation (Rs.)			Net income (Rs.)
	Wheat*	Paddy**	Wheat	Paddy	Total			Crops	Trees	Total	
Grewia optiva	1606 (29)	1642 (28)	67645	33497	101142	16964	118106	85150	7810	92960	25146
Morus alba	1592 (29)	1494 (34)	67053	30478	97531	17910	115441	85150	6710	91860	23581
E. hybrid	1611 (28)	1586 (30)	67855	32354	100209	30240	130449	85150	2210	87360	43089
Control (No. trees)	2249	2275	94725	46410	141135	–	141135	85150	–	85150	53775

Source: Khybri et al., 1992

*Average of 13 crops
**Average of 8 crops
Figures in parentheses give percentage decrease in yield as compared to control

(Table 2) that growing of crops alone was more economical— requiring less investment and providing the maximum net income. Among the tree species grown in crop combination, *E. hybrid* was the most economical. Although the *Grewia optiva* and *Morus alba* practices were the least economical, their economic value could be higher because they provide, for eight to nine years, leaf fodder for cattle raising, and feed for silkworm rearing leading to value-added products.

It is explained that the reasons for decrease in yield are shade and root competition. The yields of wheat and paddy were reduced seriously when near to the tree, and gradually increased as their distance from the tree trunk increased (Figure 6). This interaction has been shown to be related to soil moisture.

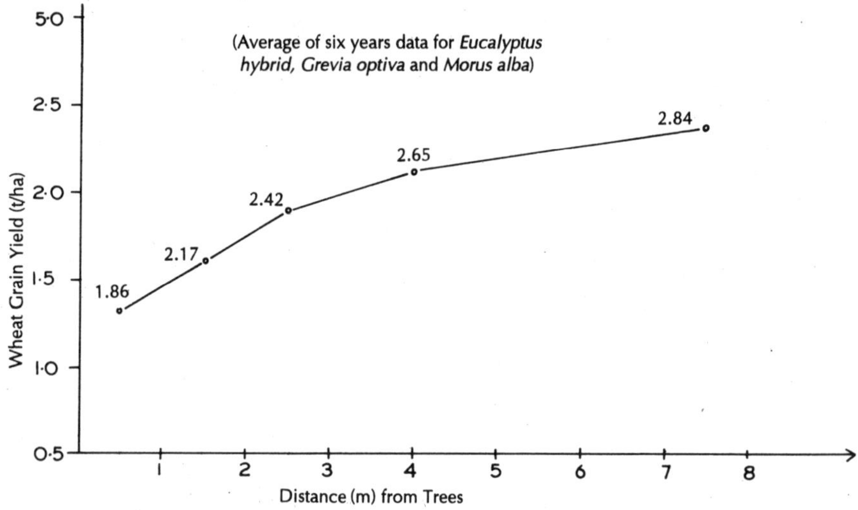

Fig. 6: Average wheat grain yield as influenced by distance of wheat crop from trees (Source: Khybri and Sewa Ram, 1984)

Resource Input and Utilisation

Apart from land, the only other input needed is planting material and labour. Labour is invariably provided by the farmer and his family.

Yields

The average yield of green leaf fodder from a fully mature *Grewia optiva* tree is 25-30 kg; taking 30 trees as the average number of trees per farm holding of one hectare, the average yield of fodder will be about 0.80 tonne per annum. Leaf fodder is not sold but utilised for the farm cattle. No data are available for the yield of fuelwood, poles, timber and honey.

Even in scientific investigations, currently, neither the agricultural scientists nor the foresters, take into account the income generated by farmers by growing fuelwood and fodder trees on their farms. However, the returns from livestock are reported, which could be an indirect indication of the returns from fodder obtained from within the farm and from outside the farm (village common land and forest land). For example, it is reported that in a watershed of 369 ha in Garhwal, UP with 94 households, milk production accounted for 35% of the total income, and livestock sales accounted for 11% of the total income. It may be noted that these values do not take into account the animal power used in agricultural operations (CSWCRTI, 1981).

Protective and Service Aspects

There are no quantitative data available on soil improvement, productivity maintenance, conservation, hydrological cycle/catchment effects etc. Presumably trees provide protection against soil erosion. With the recent advances in agricultural technology, and emphasis on promotion of farm forestry, progressive farmers are increasingly adopting improved practices and better management. These agroforestry practices, which are traditional, presumably improve overall production of the farm, socio-economic status of the farmer, better maintenance and improvement of the site etc. Farms on which agroforestry is not practised are comparatively poorer in every respect.

Improvement of the Practice

In the past, the farmers used to tend only self-sown seedlings of tree species which they considered useful for fodder and fuelwood, and the bulk of their fodder and fuel requirements were met from nearby forests. With the steady recession of forests, farmers now make a conscious effort to procure/purchase nursery-raised plants. Some of the aspects of this practice, which need investigation, are choice of species, selection of fast-growing and high-yielding tree species, frequency and intensity of lopping of tree species, interaction between trees and crops, and input and output relationships.

2.1.3 *Acacia* Species

Many *Acacia* species are grown extensively in and around agricultural fields throughout India. Despite this, and surprisingly, there is no detailed information about this practice. Perhaps, the presence of *Acacia* species is taken for granted.

Acacia nilotica spp. *indica*, grows in dry to moist tropical climate. It is allowed to grow in and around agricultural fields all over India. Almost every part of the tree is used. Its leaves, young twigs and pods are used as fodder; the branches are used as fuelwood and fencing material; the bark is used for the extraction of tannin; the timber is used for making agricultural implements; and the tree yields gum.

It is reported that in the Punjab Shiwalik region farmers grow an average of 22 trees per hectare. *Acacia nilotica* was found to be the most dominant (83.3%) (Grewal, 1992). It is further reported that in the semi-arid region in Rajasthan, agricultural crops are grown in association with *A. nilotica* (var *cuppressiformis* and *A. nilotica* var *radiane*); with *Salvadora persica* and *S. oleoides* (Malhotra, 1984).

In semi-arid and arid conditions, since there is insufficient soil moisture even for crop growth, *A. nilotica* interferes severely with crop growth. In addition, its dense canopy and shade also affect crop growth adversely. The yield of irrigated wheat crop is also reported to be significantly less (42%) up to 4 m distance from *A. nilotica*. The yield of wheat increased gradually 4-15 m away from the trees; it was significantly higher when the distance from the tress was more than 15 m (Sharma, 1992). Dhillon et al. (1984) also reported 40-94% reduction in the yield of wheat grown up to a distance of 7 m from the *A. nilotica* tree line. However, it would appear that *A. nilotica* gives good returns to the farmers. It is reported (Grewal, 1992) that apart from the annual yield of leaf fodder, pods and fuelwood, farmers sell trees regularly. In one case four farmers sold 67 trees to contractors and generated an income of Rs. 13,800.

Acacia nilotica is allowed to grow in paddy fields in the sub-humid sub-tropical districts of Orissa. However, its impact—adverse or otherwise —on the paddy yield is not reported.

Acacia nilotica trees are generally mismanaged. Sinha (1958) suggested that for proper lopping management, the upper one-third of the crown, and branches above 7.5 cm in diameter in the lower one-third of the crown should not be lopped, and adequate time gap should be provided between consecutive loppings.

Acacia leucophloea

Shanmugasundaram (1984) reported that *A. leucophloea* is grown in the dry districts of Coimbatore, Tiruchi and Madurai in Tamil Nadu. Jambulingam and Fernandes (1986) reported that it is grown in over 100,000 ha in the dry tracts of Coimbatore and Periyar districts of Tamil Nadu. The rainfall is erratic from year to year.

Acacia leucophloea regenerates profusely when the land is ploughed after the first rain. Farmers encourage the growth of young seedlings and also sow crops like *Pennisetum glaucum* and *Dolichos biflorus* (horse gram). Depending upon rainfall, the crops are harvested for grain or fodder. At the end of the first year, about 1,000 tree seedlings remain; this number is further reduced by natural mortality to around 500-800/ha at the end of the third year, and the farmer continues to till the soil up to the base of the trees to sow the crops. After about 10 years, when the trees are around 10 m tall and 20 cm diameter at breast height, they are thinned to 60-100 trees/ha. It is reported that about 20% more yield of sorghum dry fodder is obtained when grown in association with *Acacia leucophloea* (Sheshadri, 1976). The trees develop a large spreading crown when 15-20 years old, and at that age they are thinned out to 25-60 trees/ha. Such trees yield up to 100 kg pods annually, which form an excellent high-protein fodder supplement in the dry season. Its wood is used for timber turning and fuel. Due to the uncertainty of rains and other socio-economic reasons, an increasing number of farmers are now planting *Cenchrus ciliaris* grass instead of cereals and pulses, thus the practice is changing from agrosilviculture to pastoralsilviculture. *Acacia leucophloea* is grown on farm lands and field bunds in Orissa state also (Das and Mahapatra, 1984).

Acacia planifrons

Jambulingam and Fernandes (1986) reported that *Acacia planifrons* grows extensively in the coastal parts of Tirunelveli and in localised pockets of Ramanathapuram and Coimbatore districts in Tamil Nadu. Regeneration of this species is achieved by penning on agricultural lands, goats fed on the pods of this species. In order to ensure good regeneration (up to 80%) the land is ploughed prior to penning. At the end of one year, the population is thinned down to 1,250-1,500 seedlings/ha. Sorghum, pulses, groundnut and chillies are grown among the trees for four to five years until the tree canopy closes. The trees are felled between the seventh and tenth years depending upon the socio-economic condition of the farmers. Trees provide wood for fuel, and valuable fodder pods during the dry season.

Trees of *Acacia catechu, Acacia modesta, Acacia nilotica, Dalbergia sissoo, Morus alba* and other species, which regenerate naturally in the fields, are retained by the farmers of the Punjab in the foothills of the Shivaliks (Dhanda, 1984; Grewal, 1992).

2.1.4 Other Tree Species

India has a wide variety of agro-ecological regions and therefore a great variety of trees grow all over. Farmers have integrated trees with various farming systems. Two classic traditional examples, one for the arid and semi-arid zone in the Rajasthan desert and the other for the sub-tropical and temperate zone in the Himalaya, and some details about *Acacia* spp. are described above. Apart from these, many other species are grown in the agricultural fields, but unfortunately detailed descriptions and quantitive data are not available, since neither are they recognised as a component of crop farming by agricultural scientists, nor are they recognised as a component of forestry by foresters. Therefore, the practices remain neglected. Some useful information, although still incomplete, is however emerging.

Borassus flabellifer

It is conservatively estimated that 40 million *Borassus flabellifer* (palmyrah palm) trees are growing in the plains of Tamil Nadu (Jambulingam and Fernandes, 1986). It is grown in Andhra Pradesh also (Rao 1984). Palmyrah palm trees are dispersed on the cropland as well as on bunds of fields. The adverse effects of shading caused by this palm on the crops are likely to be negligible because of its small crown and great height. No yield data of crops are reported. Reduction in crops yield, if any, due to root competition, is surmised to be compensated for by the sale of various products obtained from the palm, and cash income during the dry season when there is no other crop in the fields. Major products of the palm are *nera* (a sugar exudate), leaves (used for thatching and making handicrafts), fibre (from petioles), timber and fuelwood. *Nera* is tapped daily for five months in a year (January to May) and a single tree may yield about 2 litres a day in the northern dry districts and 2.5-3 litres a day in the southern wet districts of Tamil Nadu. Tapping starts when a tree is 14 to 16 years old. *Nera* is used as a drink and converted into palm sugar. The total annual income from the sale of palmyrah products such as palm sugar, candy and handicrafts was estimated at Rs. 120 million in 1986. About 28,000 families are engaged in the processing of palmyrah products (Jambulingam and Fernandes, 1986). This tree, like the coconut, is also known as *Kalpa Vriksha* (a mythological term, implying an object which provides for most human needs), in the villages of Tamil Nadu. It can be grown in poor and sandy soils and is an ideal species for the semi-arid areas.

Tamarindus indica

The tamarind tree occurs throughout Tamil Nadu. It is planted at 8 × 8 m, 8 × 12 m and 12 × 12 m spacing. Reduction in agricultural crop yields caused by this tree is reported to be severe due to the shading effect of the dense and large crowns of mature trees. However, most farmers plant the trees on poorer sites. A mature tree (20 years) can yield about 100 to 500 kg pods per year giving 50-250 kg of edible pulp. Usually the pod yield is cyclic with bumper yields every third year. The seed is ground, and the starch obtained from the cotyledons is processed to give a gum which is used for sizing textiles, paper and jute. Once the pod yield starts declining (50 years), the trees are cut for fuelwood and charcoal (Jambulingam and Fernandes, 1986). Tamarind is grown by farmers on their fields in Andhra Pradesh also (Rao, 1984). In recent years some high-yielding tamarind trees have been selected which are being propagated by vegetative means.

Ceiba pentandra

Kapok trees are planted at 5 × 5 m or 8 × 8 m spacing, and intercropped with cereals or pulses in Tamil Nadu, till the canopy closes in about seven to eight years (Jambulingam and Fernandes, 1986). Yields from interplanted crops can be expected to decline from the fourth to the seventh year after planting the trees. Trees start fruiting from the third or fourth year onwards; stabilised kapok yields are obtained from the seventh or eighth year onwards. A well-managed 15-year-old tree can yield 1,000-2,000 pods. The small variety of this tree yields about 6 kg floss, and the Singapore kapok variety yields 7-9 kg floss per 1,000 pods. A farmer can obtain an annual return of at least Rs. 120 per tree. Its wood is used for making matches and a 20-year-old tree can fetch Rs. 300 to 500. The trees shed all the leaves in the dry season, thus helping in rehabilitating poor or degraded farmland.

Leucaena leucocephala

Of late a great deal of attention has been paid to *Leucaena leucocephala* (*subabul*) almost in the whole of India, as a multipurpose fodder, fuelwood, timber, nitrogen fixing, and very fast-growing tree, and much benefit is being attributed to it. While, it is being popularised on its assumed good qualities, not much factual information is available about its interaction with agricultural crops. Some experimental data indicate that yield of *subabul* fodder and fuelwood is influenced by rainfall. Under rainfed conditions, it seems to yield more fodder and fuelwood in the 1,000-1,500 mm annual rainfall range than in high (1,500-2,000 mm) or low rainfall (500-750 mm) (Table 3). Further, and as it should be, it yields more fodder when it is closely spaced (75 × 25 cm) than when widely spaced (375 × 25 cm); when it is intercropped, its fodder and fuelwood yields generally tend to go down, but there is an indication that fertilisers applied to crops contribute to its growth; the intercrops yield less than pure crops; the yield and type of crop grown is determined by the amount of rainfall (Khybri et al., 1988; Grewal et al., 1992; Prasad and Verma, 1987).

Table 3: Intercropping with *Leucaena leucocephala*

	Place			Remarks
	Dehradun	Chandigarh	Kota	
Annual rainfall (mm)	1500–2000	1000–1500	500–750	
Yield of fodder *(tonnes/ha)*				
Pure *subabul* (75 × 25cm)	9.4 (airdry)*	34.0 (green)**	37.3 (green)a	* Av. of 7 years 1981–88. Planted in 1981.
Pure *subabul* (375 × 25cm)	5.3 (airdry)*	18.9 (green)**	14.8 (green)a	** Av. of 6 years 1981–87. Planted in 1981.
Intercropped (375 × 25cm)	5.3 (airdry)*	16.4 (green)***	8.9 (green)a	*** Av. of 6 years, 1981–87. Planted in 1981 with three different crops—maize, black gram, cluster beans.
Yield of fuelwood *(tonnes/ha) (air dry)*				
Pure *subabul* (75 × 25cm)	6.9 b	9.4 c	–	a. Av. of 7 years, 1981–87. Planted in 1981.
Pure *subabul* (375 × 25cm)	3.3 b	6.3 c	–	b. Av. of 5 years 1983–88. Planted in 1981.
Intercropped (75 × 25cm)	3.5 b	5.4 c	–	c. Av. of 4 years 1983–87. Planted in 1981.
				d. Av. of 6 years 1981–87.
				e. Av. of 6 years 1981–87.
				f. Av. of 7 years 1981–87. (6 crops)
Yield of intercrops *(kg/ha)*				
Wheat grain	1970 d			
Wheat straw	3255 d			
Paddy grain	1590 d			
Paddy straw	3663 d			

contd.

Table 3: contd.

	Place			Remarks
	Dehradun	Chandigarh	Kota	
Maize grain		1440 e		
Cluster bean grain		868 e		
Black gram grain		249 e		
Sorghum grain			1296 f	
Pigeon pea grain			564 f	
+			+	
Sorghum grain			1027 f	
Pigeon pea grain			724 f	
+			+	
Black gram grain			205 f	
Yield of pure crops (kg/ha)				
Wheat grain	2450 d			
Wheat straw	4086 d			
Paddy grain	1950 d			
Paddy straw	4438 d			
Maize grain		2296 e		
Black gram grain		403 e		
Cluster bean grain		1224 e		
Sorghum grain			1501 f	
Pigeon pea grain			689 f	
+			+	
Sorghum grain			1189 f	
Pigeon pea grain			824 f	
+			+	
Black gram			252 f	

Source: Khybri et al., 1988; Grewal et al., 1992; Prasad and Verma, 1987.

The economics of these practices at Chandigarh reported by Grewal et al. (1992), indicate that *Leucaena* grown alone at spacing of 375 × 25 cm was significantly less economical than that grown at a spacing of 75 × 25 cm. Among the three agricultural crops, black gram grown alone was the least economical. The net cash returns from the three combinations of *Leucaena* with crops, and the pure *Leucaena* at spacing of 75 m × 25 cm, were comparable and not significantly different (Table 4). Similar economics of practices are reported from Dehradun (Khybri et al., 1992) with the difference that pure cropping of paddy and wheat was much less economical than intercropping with *Leucaena* (Table 4).

Alley cropping with *L. leucocephala* has been extensively experimented with during the recent years in the semi-arid tropics in India (Singh et al., 1987c; Rao et al., 1991). It is observed that as the distance between the rows of *L. leucocephala* increased, the per cent reduction in crop yield decreased (Table 5). Rao et al. (1991) also observed the same trend between 3 m and 5.4 m wide alleys of *Leucaena* in the shallow alfisols of the

Table 4: Economics of intercropping with *Leucaena leucocephala*

Treatment	Net return (Rs. per ha per annum)	
	Location	
	Dehradun*	Chandigarh**
Leucaena leucocephala (75 × 25 cm)	5316	3128
Leucaena leucocaphala (375 × 25 cm)	2553	2360
Leucaena leucocephala (375 × 25 cm) intercropped with		
– Wheat & paddy	5355	
– Maize		3273
– Blackgram		3125
– Cluster bean		3540
Pure crops		
– Wheat & paddy	2727	
– Maize		2337
– Blackgram		1559
– Cluster bean		3013
L.S.D.(P. 0.05)	–	597

Source: Khybri et al., 1992*; Grewal et al., 1992**

Average of: six years of crops, seven years of green fodder and five years of fuelwood at Dehradun;
six years of crops, six years of green fodder and four years of fuelwood at Chandigarh.

Table 5: Effect of row spacing of *Leucaena leucocephala* on yield of pearl millet

Location	Tree row spacing (m)	Seasonal rainfall (mm)	Grain yield (kg/ha)		Per cent reduction in yield
			Sole crop	Alley crop	
Anantpur	2	326	240	100	58
	4	326	240	180	25
Kovilpatti	3	396	730	590	19
	6	396	730	650	11
Hyderabad	3.6	190	1310	890	32
	7.8	190	1310	1220	7

Source: Singh et al., 1987c

semi-arid tropics. It is suggested that in the semi-arid tropical areas in India, tree rows should be 10, 15 or 20 m apart. This almost amounts to accepting the traditional practice of the farmers to have trees on field boundaries or field bunds which are set far apart.

With respect to the age of the alley crop, it is observed that the yield of the crop decreased consistently and progressively as the age of the tree increased (Singh et al., 1987c). The reduction in the crop yield in the case of wider spacing (7.8 m) was relatively less for the same age of the tree as compared to the 3.6 m row spacing. This wider row spacing, to some extent, is able to mitigate the adverse impact of the greater age of the tree.

With respect to the response of different crops, castor and pigeon pea were more adversely affected by the presence of *L. leucocephala* than sorghum and pearl millet (Table 6). Rao et al. (1991) also reported severe reduction in the yield of pearl millet, pigeon pea, castor and groundnut due to alley cropping.

Table 6: Impact of *Leucaena leucocephala* on yield of different intercrops

Crop	Year						Per cent mean reduction
	1984		1985		1986		
	Grain yield kg/ha)	% reduction	Grain yield (kg/ha)	% reduction	Grain yield (kg/ha)	% reduction	
Sorghum	548	49	636	68	909	11	48
Pearl millet	720	53	997	51	2026	+14	30
Pigeon pea	227	61	9	65	–	–	61
Castor	284	56	149	77	211	47	62
***Rainfall during crop growth** (mm)*	1984		1985		1986		
Sorghum & pearl millet	208.6		145.1		280.0		
Castor & pigeon pea	329.4		286.9		300.9		

Source: Singh et al., 1987c

Note: + 14 indicates increase in pearl millet yield

Leucaena leucocephala being a multiple-purpose tree can be utilised for mulch, fodder, fuelwood and poles singly or in various combinations. Attempts have been made to manage *Leucaena* trees planted at 5 m row spacing and allowing *Leucaena* plants at every 2 m spacing within the row to grow into a pole; *Leucaena* rows were lopped at 1 m height every two months during the crop season, and at every four months during the no-crop season. While crop yields are reported, the yield/income of poles is not reported. In fact all the above experiments suffer from a common weakness of the non-availability of yield of the tree component and economics of various treatment combinations. Singh et al. (1987c) recommend alley cropping on the basis of reducing the risk factor, since *Leucaena* trees yield some green fodder during the years when the crops fail in the drought years (Table 7). Rao et al. (1991) have reported the economics of all the components of alley cropping. On the basis of four years data, annual crops without mulch gave better economic returns (Rs. 3,515/ha/annum) than alley cropping without mulch at 3 m (Rs. 2,744/ha/annum) and at 5.4 m (Rs. 3,291/ha/annum). Alley cropping at 5.4 m was more economical (Rs. 3,291 and Rs. 599/ha/annum without and with mulch, respectively) than at 3 m (Rs. 2,744 and Rs. 980/ha/annum without and with mulch, respectively). Mulching of annual crops (Rs. 461/ha/annum), and alley cropping was uneconomical in all cases as compared to no mulch. A sole crop of *L. leucocephala* (Rs. 2,552/ha/annum) was less economical than annual crops without mulch. Only alley cropping at 5.4 m without mulch was more economical than sole *L. leucocephala*. *Leucaena* by itself was more economical in years when there was moisture stress for crops.

Table 7: Risk reduction in drought years through alley cropping at Rajkot

Treatment	*Leucaena* fodder (wet) (t/ha)	Crop yield (t/ha)	
		Grain	Stover
Sole crops			
*Leucaena**	5.75	–	–
Groundnut	–	0	1.67
Green gram	–	0	0.68
Black gram	–	0	0.47
Alley crops			
Groundnut	5.24	0	1.26
Green gram	5.56	0	0.25
Black gram	6.03	0	0.22

Source: Singh et al., 1987

*: 1.5 m rows

Note: Rainfall 177 mm (120 mm in cropping season) as against 625 mm average.

Prosopis juliflora (mesquite)

This leguminous tree was introduced in the semi-arid Kurnool district (now in Andhra Pradesh) in the late nineteenth century. It now grows wild in large areas in Andhra Pradesh, Karnataka, Kerala and Tamil Nadu. Farmers in the Ramanathapuram and Tirunelveli districts of Tamil Nadu grow mesquite to reclaim wastelands and saline soils for agricultural crops

(Shanmugasundaram, 1984; Jambulingam and Fernandes, 1986). In Ramanathapuram district of Tamil Nadu, farmers use *Prosopis juliflora* as a fallow species for four years. This ameliorates the soil sufficiently to enable them to grow annual food crops for at least two years, after which the trees are allowed to reinvade the site for another four years. The fuelwood produced by this tree is of good quality and is excellent for making charcoal. It is a major source of fuelwood and income for many rural people in Andhra Pradesh and Tamil Nadu. If cut under the coppice system once in five years it yields 50 tonnes of fuelwood per hectare (Shanmugasundaram, 1984).

For a long time this tree was considered a weed by foresters as well as farmers, but not any more. It has been lately realised that while contrary to expectation, the community woodlots have provided very little fuelwood for the rural people, *P. juliflora* has been meeting this need to the extent of 90% in Tamil Nadu. In addition, it is an important source of income for the rural poor who sell fuelwood and charcoal to towns and urban areas. No permit is required to collect, transport and sell the fuelwood.

Farmers grow/retain *Azadirachta indica, Pongamia pinnata, mango* (*Mangifera indica*), *Madhuca latifolia* etc. in their fields in Andhra Pradesh (Rao, 1984); farmers raise bamboos in Gujarat on farms, and near river banks, subject to soil erosion (Unadkat, 1984).

Farmers in the foothills of the Shiwaliks (Kandi Area) in the Punjab retain trees of *Acacia catechu, Acacia nilotica, Butea monosperma, Dalbergia sissoo, Mangifera indica, Acacia modesta* and *Zizyphus mauratiana* on their farms wherever they happen to regenerate naturally (Dhanda, 1984; Grewal, 1992).

Many other tree species are now being tried in agricultural fields. For example *Sesbania sesban* with rainfed crops of sesamum, groundnut and pigeon pea; *S. sesban* and *S. grandiflora* with irrigated wheat, sorghum (for fodder) and pearl millet (for fodder). *Sesbanias* are grown for leaf and fuelwood. They need to be defoliated and managed for better crop growth (Patil et al., 1981); *Gmelina arborea* in Gujarat (Unadkat, 1984). Paddy-wheat rotation is a very common and highly productive crop rotation in India. Paddy and wheat have very different growth requirements. Paddy needs flooding and wheat is a dryland crop. In the subhumid Terai region of UP, paddy crop gets surface flooding during the rainy season and the water level drops 1 to 2 m below the surface during winter. Some tree species have been evaluated to select a multipurpose tree suited to these edaphic conditions. *Trewia nudiflora, Dalbergia sissoo* and *Bombax ceiba* performed better than *Melia azedarach, Terminalia arjuna, Syzygium cuminii, Cassia_fistula* and *Leucaena leucocephala*. After three years of growth the three promising species had to be cut back by as much as 65 % to reduce competition for light and nutrients for the crop. The three tree species survived and grew back well.

Of the three tree species *Trewia nudiflora* was the most promising. However, the paddy yields were still adversely affected. These tree species were grown at 5 × 4 m spacing. The results indicated that wider tree spacing is desirable (Saini and Misra, 1992). *Trewia nudiflora* is one of the lesser known multipurpose tree species even though farmers are traditionally familiar with it and use it in several ways. Its fruit is used as a vegetable, its roots and shoots have medicinal value, its leaves provide a nutritious fodder (12.6% crude protein), and the seeds yield oil. The trees produce a soft timber which is used traditionally for packing cases and tea chests. The male trees have a narrow crown; most of the female trees have a spreading crown but a few have a narrow crown. There is therefore need for further work on this tree.

Farmers in Manipur grow *Morus alba, Artrocarpus chaplasha, Arundinaria, Bambusa, Dendrocalamus, Bauhinia, Ficus* species, *Trewia nudiflora* and *Alnus nepalensis* for fodder on their farms (Sarma, 1984). Fruit-bearing trees like *Mangifera indica, Syzygium cuminii, Emblica officinalis,* and *Parkia roxburghii* are planted in their homesteads.

Farmers in Orissa traditionally grow trees on farm lands and field bunds to meet their fuelwood and small timber needs.

2.2 TREES ON FARM BOUNDARIES

Trees, which are grown in agricultural fields or on field bunds, are also often and usually grown on farm boundaries. Apart from the specific uses to which these trees are put as described above, they also help in the demarcation of the farm and field boundaries, and serve as wind breaks and shelter belts. This system is very extensively practised in all the agro-ecological regions of India.

2.2.1 *Eucalyptus* Species

Among the trees which are grown on farm boundaries, *Eucalyptus* species are the most prominent and popular.

Eucalyptus tereticornis

For the past few years *Eucalyptus* species, particularly *E. tereticornis* or *E. hybrid* is being grown very extensively in many parts of India, mostly on boundaries and bunds of farms. The adverse effect of *E. tereticornis* on the yield of rainfed *rabi* wheat is already reported in table 2. *Eucalyptus citriodora* is reported to have reduced the yield of *kharif* rainfed crops of green gram and sorghum. It is reported (Tomar and Shrivastava, 1986) that when *Eucalyptus* was grown on the bunds of paddy fields, there was reduction in the incidence of light by 63.3% at 0900 hours, 1.7% at noon, and increase by 3.8% at 1600 hours; on an average the paddy plant received 24.2% less light during seven hours (Table 8). The growth of a paddy plant as measured by plant height, number of tillers and leaves per plant was adversely affected by the shade, and hence the yield was reduced by 24.6%. It is also surmised that the adverse effect of *Eucalyptus* on the paddy plant may not be entirely due to shade but may also be due to the competition for nutrients and moisture.

The crown of the mango tree, as compared to the *Eucalyptus*, forms a dense shade and paddy plant therefore receives much less light. As a consequence the paddy plants were much more adversely affected and the crop yield was reduced by 55.5% in the shade (Table 8).

Table 8: Effect of *Eucalyptus* and mango trees on growth and yield of paddy

Parameter	Open	Shade	% reduction due to shade
	Eucalyptus–Paddy		
*Light intensity (Lux)**			
0900 hrs	1467	540	63.3
1200 hrs	1900	1967	1.7
1600 hrs	520	540	+ 3.8
Average for 7 hours	1296	982	24.2
Plant height (cm)**	104.4	99.3	–
Effective tillers per plant (number)	13.0	9.6	–
Non-effective tillers per plant (number)	0.5	0.6	–
Leaves per plant (number)	63.6	49.6	–
Yield (kg/ha)	5196	3916	24.6
	Mango–Paddy		
Light intensity (Lux)			
0900 hrs	1400	176	87.3
1200 hrs	1933	733	61.1
1600 hrs	517	333	35.4
Average for 7 hours	1283	414	67.7
Plant height (cm)	993	83.2	–
Effective tillers per plant (number)	11.3	8.1	–
Non-effective tillers per plant (number)	1.0	0.2	–
Leaves per plant (number)	57.4	41.1	–
Yield (kg/ha)	4875	2170	55.5

Source: Tomar and Shrivastava, 1986
*Light intensity is measured on top of the rice plants
**Plant parameters refer to paddy plant

In a well-documented experiment in which *Eucalyptus tereticornis* and *Leucaena leucocephala* were grown on the field boundaries at a plant to plant distance of 2 m, along 40 × 20 m plots, and irrigated crops were grown for five years, it was observed (Table 9) that the yield of irrigated tobacco and pearl millet grown in rotation was adversely affected by trees grown on the field boundaries. It was also found that *Eucalyptus* had a greater adverse impact than *Leucaena*, and that *Eucalyptus* affected the crop yield adversely over a longer distance than *Leucaena*.

Table 9: Effect of *Eucalyptus tereticornis* and *Leucaena leucocephala* grown on field boundaries on yield of irrigated crops

Yield/Economic parameters	Parameters (No trees)	Treatment	
		Eucalyptus on field boundary	*Leucaena* on field boundary
Yield (kg/ha)			
Tobacco			
1985	2210	2250	2410
1986	1660	1540 (–7.2)	1800 (8.4)
1987	1120	610 (–45.5)	930 (–16.7)
1988	700	330 (–57.7)	590 (–24.3)
1989	–	– (–58.3)	– (–37.1)
Pearl Millet			
1985	1750	2090	2090
1986	2130	1680 (–21.0)	2190 (2.8)
1987	1840	1270 (–31.0)	1860 (1.0)
1988	2110	790 (–62.5)	1760 (–16.6)
1989	–	– (–36.9)	– (–13.8)
Distance (m) of adverse impact of tree line on crop yield			
Tobacco			
1987		4	
1988		6	3.8
Pearl Millet			
1987		4	
1988		6	2.5

Source: Nambiar et al., 1986, 1987, 1988
Nambiar and Singh, 1990
Figures in parentheses indicate percentage decrease or increase in yield due to the treatment.

As the economic analysis of the entire practice described above has not been reported, it is difficult to assess whether agricultural cropping or agroforestry is more profitable. However, an economic evaluation of the trees grown on boundaries indicated that *Eucalyptus tereticornis* was much more profitable than *Leucaena leucocephala*. This is partly due to the fact that poles of *Eucalyptus* are much valued, and also due to the fact that *Leucaena* is prone to damage by wild animals (Table 10). It is not clear whether a net income of Rs. 20,546 over a period of six years from the trees would compensate for the loss of cash income from tobacco and pearl millet over a five-year period of cropping.

It is reported that *Eucalyptus* and other trees grown on the boundaries of irrigated fields decreased crop yields; the extent of reduction of the crop

Table 10: Growth parameters, yield and income from *Eucalyptus tereticornis* and *Leucaena leucocephala*—grown on irrigated field boundaries (200 trees/ha)

Observation parameters	*Eucalyptus*	*Laucaena**
Average survival (%)	92.1	70.4
Average dbh (cm) in 6 years	16.2	9.9
Average height (m) in 6 years	15.7	8.7
Number of poles obtained/ha (length 6 m)	289	0**
Total airdry fuelwood obtained in 6 years (kg)	7531	13589
Total airdry lops & tops obtained in 6 years (kg)	7485	3023
Total green fodder obtained in 6 years (kg)	–	3971
Income from poles at the rate of Rs. 65/pole (Rs.)	18785	–
Income from fuelwood at the rate of Rs. 8/20 kg (Rs.)	3012	5436
Income from lops & tops @ Rs. 2.20 kg (Rs.)	749	302
Income from green fodder at the rate of Rs. 2.20/kg (Rs.)	–	397
Gross income (Rs./ha)	22546	6135
Total cost of raising, maintenance & harvest of trees/ha at the rate Rs. 10/tree (Rs.)	2000	2000
Net income in 6 years (Rs./ha)	20546	4135

Source: Nambiar and Singh, 1991.

*The growth of *Leucaena* was vitiated due to damage caused by blue bulls and monkeys. Since the species is highly palatable, it needs protection from all biotic interferences for getting normal growth and form.

**The growth form was vitiated due to the above reason and therefore there were no normal straight poles and hence only fuelwood material was obtained.

yield is said to be influenced by the direction of the tree lines. When the tree line was on the northern aspect, 1.6 to 64.6% reduction in the yield of wheat was recorded; this was 0.3 to 36.5% when the tree line was either on the eastern or western aspects; there was no reduction in the yield when the tree line was on the southern aspect (Dhillon et al., 1979, 1982). The reduction in yield was dependent on the tree species, height, crown and spacing.

Eucalyptus globulus

This tree was introduced in 1843 (Samraj, 1977) from Tasmania in the Nilgiri Hills; it is now grown in high rainfall areas, on the high hills of South India, on field boundaries and terrace risers. The trees are grown primarily for harvesting leaf to extract *Eucalyptus* oil. No data on the economics of growing these trees in an agrosilviculture practice are available. It is reported that the yield of potato is adversely affected by growing *E. globulus* on bench terrace risers (21.0 tonnes potato/ha when

E. globulus is not grown and 17.5 tonnes potato/ha when *E. globulus* is grown) (Seshachalam, 1979; Tejwani, 1981). It is obvious that the farmer finds it economical to sacrifice some yield of potato to grow *E. globulus* for its oil and fuelwood. In an experiment, where *E. globulus* was planted at 2 × 2 m spacing and potato was grown as an intercrop (*taungya* system), the yield of potato declined severely in the second year of plantation. The trend was the same when potato was grown with *Acacia mearsnii*. Growth parameters of *E. globulus* were much better when potato was grown as an intercrop than when geranium, or grass + legume mixture were the intercrops. This was due to the fact that *E. globulus* benefited from the fertilisers applied to potato (Samraj et al., 1982).

It is reported that the tap root of a mature tree of *Eucalyptus globulus* penetrates as much as 6 m deep in the ground, and its superficial lateral roots spread out to about 8 m length. The superficial roots reduce the yield of crops to the tune of 30 to 40%. It has been observed that about 75% of the lateral roots can be prevented from invading the crop root zone by planting tall seedlings in pits 60 × 75 × 90 cm in case trees are to be grown on farm boundaries (Jeevarathnam et al., 1992). In case of woodlots near agricultural fields a trench 75 × 45 × 90 cm is recommended.

It could be reasonably concluded that all the *Eucalyptus species.* being grown in India, adversely affect the yield of agricultural crops, but it is not known as to whether the income obtainable from *Eucalyptus* trees compensates for the loss in the yield of the agricultural crops grown.

The price obtained from *Eucalyptus* trees, grown on farm boundaries depends, among other things, on the age and size of the tree, and nearness to the market. It is reported that the prices of *Eucalyptus* progressively increased from 1976 onwards in Haryana (Hooda, 1984). In Gujarat, with 80 and 60% survival and at 10% interest, the benefit : cost ratio of 3.1 and 2.71 respectively was obtained, without taking into account the adverse impact of trees on the yield of crops (Malwade, 1984). However, since 1986 there has been a steep fall in the prices of *Eucalyptus* due to various reasons, the chief reason being the arrival of large stocks in the market.

2.2.2 *Leuceana leucocephala*

While *Leuceana leucocephala* is not new to India, its plantation received a great boost due to very intense 'promotion' efforts eulogising its multipurpose qualities of fast growth; its use as fodder, fuelwood, and small wood; its nitrogen-fixing and leaf-manure value. Even though these values were not tested in India, but elsewhere, the tree was vigorously promoted as a cure-all 'wonder' tree. Its fodder value, however, has not been without controversy since it was reported from Australia that when it was continuously fed to sheep, they lost their wool.

Leucaena leucocephala planted along farm boundaries in Bundelkhand region is reported to yield 3.21 tonnes fuelwood and 1.44 tonnes forage per ha three years after planting. (Patil et al., 1981). The authors do not mention the spacing or number of plants per hectare. It is stated that the tree does not interfere with cropping. The experimental results on growing this tree in agricultural fields are reported in section 2.1.4.

2.2.3 Other Tree Species

Delonix elata, a small evergreen, leguminous tree, capable of giving green manure is very popular with farmers in Tamil Nadu, where they grow it on field bunds and boundaries of irrigated paddy fields. *Tectona grandis* is very popular with the farmers of Thanjavur district in Tamil Nadu (Srimathi and Venkatesan, 1981). *Syzygium cuminii* and *Glyricidia* spp. are also grown for green manure on field bunds in wet paddy lands in Tamil Nadu (Nair, 1981).

Prosopis juliflora when grown on farm boundaries, is observed to very adversely affect the yield of agricultural crops. Prajapati et al. (1971) reported that the yield of sorghum was adversely affected up to a distance of 30 m from the tree (four to five times the height of the tree). On a large farm, where the fence is separated from the field crops by a road, *P. juliflora,* is reported to yield about 2,700, 2,700, 2,400 and 2,500 kg total air-dry mass per kilometre length of the fence per year, with 6, 12, 18 and 24 months lopping cycle, respectively (Subbayan, 1984). Yield of about 1,000 kg of air dry smallwood (between 5 and 20 cm diameter) and about 1,500 kg of air dry twigs (diameter less than 5 cm) per km length of fence per year was obtained when the trees were lopped at 12 and 18 months cycle.

Dalbergia sissoo when planted in the semi-arid areas of Haryana under the social forestry programme, served as an effective wind break for crop production. The beneficial impact of 18-year-old and 18 m tall trees of *Dalbergia sissoo* grown on field boundaries, was reflected in decrease in wind velocity by 15 to 45% depending on the season and wind speed; and in general, the growth and productivity of irrigated cotton increased up to four times the tree height towards the leeward side. Depending on the orientation of the tree belt, the yield of cotton increased by 4-10% (Puri et al., 1992).

Willows and poplars are grown along field boundaries in Lahaul in Himachal Pradesh.

Farmers in Sikkim grow bamboo (*Dendrocalamus hamiltonii,* and *D. sikkimensis*) in agricultural fields all along the irrigation channels and stream banks to meet the fodder needs of their livestock—usually one or two milch cows and three to four goats—in the winter months. Bamboos are extensively used in building small farmhouses, goat sheds and piggery enclosures; large bamboo pipes are used as water conveyors on farm irrigation systems; bamboos are used for making baskets. Farmers in Sikkim also grow a variety of tree fodders, of which the *Ficus* species (*Ficus bengalensis, F. cunia, F. elastica, F. clavata, F. hirta, F. hookerii, F. infectoria,* and *F. nemoralis*) are more prominent (Venugopal, 1986)

Although north Bihar has very little forest, it is rich in orchards and other trees (Tripathi, 1984). This is due to the tree culture practised since long. There were several reasons for it, such as very large land holdings belonging to some people, which could not be cultivated wholly; uplands unsuited for agriculture; and domestic and local needs for fuelwood and timber. *Dalbergia sissoo* and *Wendlandia exserta* are planted on bunds made on the boundaries of agricultural fields or orchards. *Wendlandia exserta* does not yield fodder but gives poles of about 15 cm diameter after 8-10 years. *Casuarina equisetifolia* is extensively planted on field bunds and in coastal areas in Orissa (Das and Mohapatra, 1984).

2.3 WOODLOTS

In many parts of India, farmers grow trees in separate woodlots along with agricultural fields. In Andhra Pradesh the weekly *bazaar* (market) and various social, cultural and religious festivals of rural communities are held around the community woodlots (Rao, 1984). The practice of woodlots is expanding fast as the shortage of fuelwood is becoming more acute, or there is demand for poles in the urban areas and in industry, or for pulpwood. Woodlots are being raised mostly on large farms due to the increasing labour costs and problems of labour management, lack of irrigation, relatively less income from cropping, less risk in growing trees as compared to cropping. However, fundamentally the market forces determine the growth and expansion of this practice.

2.3.1 *Casuarina equisetifolia*

This tree is extensively grown in Andhra Pradesh, Tamil Nadu, Karnataka, and Orissa on lands which are too poor or not suitable for profitable agriculture. In Tamil Nadu it is grown on sandy, less fertile soils in Chengalpatta, South Arcot, Thanjavur, Puducottai and Ramanathapuram districts (Shanmugasundaram, 1984). This practice has been adopted by the farmers of Tamil Nadu much before India started to popularise farm forestry. The trees are planted very close, 0.5 × 1.0 m, 1.0 × 1.0 m, or 1.5 × 1.5 m spacing and harvested in five to seven years time when they may yield about 120 tonnes fuelwood/ha. Short gestation period, rapid growth, small inputs and the ability of the root nodules to fix atmospheric nitrogen, make *Casuarina* a popular tree with the farmers. The roots of *Casuarina* form nodules having the nitrogen-fixing bacterium—*Frankia*. It is estimated that *Casuarina* with *Frankia* and mycorrhizae can fix from 50 to 200 kg nitrogen per hectare (Pokhriyal and Maithani, 1989). In association with mycorrhizae it can also enhance the phosphorus status of the soil (Nair et al., 1984). *Casuarina* is subject to a lethal wilt disease. It is reported that this can be prevented by inocculating the nursery soil with *Frankia* and mycorrhizae inocculum (Gunjal and Patil, 1992). There was a negative correlation between the mycorrhizal colonisation of *Casuarina* and the wilt (*Trichosporium*) infection, suggesting that symbiotic mycorrhizal fungi might be good inhibitive agents for the biological control of the wilt disease of *Casuarina*. *Casuarina* is clearfelled, and stumps which also fetch a good price are uprooted. Often the land is sequentially brought under agriculture for a year or two when it will be reverted back to *Casuarina* (Reddy, 1981). Agricultural crops like groundnut, vegetables, cucumber and watermelon may be taken during the first year of tree plantation. Its internal rate of return (IRR) is estimated to be 24.6%. The practice attracts both the small and large farm holders (Shanmugasundaram, 1984).

Casuarina equisetifolia grows to a height of 15 m and girth of 85 cm in five years under irrigated, and in eight years under unirrigated conditions in Gujarat. It is being planted at 2 × 2 m spacing in degraded farmlands and grasslands under rainfed conditions in south Gujarat; at 2.5 × 1.2 m under irrigated conditions; and at 4.0 x 0.8 m spacing as an intercrop in sugar cane. It is also being grown in paddy fields and mango orchards (Unadkat, 1984). It serves as a wind break and it has been observed that mango trees in association with *Casuarina* remain free of fungal attack. Apart from other uses, branches of *Casuarina* are used to construct sheds for storing mango, thus saving money for the farmers. *Casuarina equisetifolia* is extensively planted as wind breaks in the coastal areas in Orissa (Das and Mohapatra, 1984).

2.3.2 *Eucalyptus* Species

Farmers in the Punjab, Haryana and Gujarat have started diversifying their farming practices due to the market economic pressures. It is now clear to them that due to the acute shortage of fuelwood as well as the uncertain crop prices, except for wheat and paddy (which are supported by the Government), *Eucalyptus* plantations could be commercially more viable than cropping, especially under irrigated conditions.

In Haryana, the net irrigated area is 1.9 m ha (52.5% of the total cultivated land) and growing of *Eucalyptus* hybrid as woodlots has proved to be popular and successful. The Haryana forest department recommended planting *Eucalyptus* at 3 × 3 m spacing along with agricultural crops. The farmers now plant 2,500 trees or more per hectare as pure woodlots (Hooda, 1984). Under irrigated plantation of *Eucalyptus hybrid* at 3 × 2 m, 2.5 × 2.5 m and 3 × 1.5 m spacing, reduction in yield of wheat and cotton is reported (Pillai, 1984). The yield of wheat without *Eucalyptus* varied from 2,500 to 4,000 kg/ha. The loss of wheat yield was observed to be about 500 kg/ha per year for the first four years of plantation; for the period of the fifth to the seventh years of plantation, there was a loss of 750 kg/ha/year. In the case of irrigated cotton the normal yield is 400-500 kg/ha; the loss of yield was 300 kg/ha/year for the first two years of *Eucalyptus* plantation, and 350 kg/ha/year for the third to the seventh years of plantation. Loss in the yield of crops is not only due to the growth of trees, but also due to the loss of land area to trees. In spite of these losses and paying interest at 12.5%, a net income of Rs.170,814 per hectare was estimated at the end of eight years, assuming a sale of 1,500 trees. It is reported that *Eucalyptus* trees grown on farm boundaries of irrigated fields and irrigation channels, gave twice the yield of fuelwood if allowed to grow for a period of 15 years than if allowed to grow for only 10 years. It was also observed that selling the trees as fuelwood gave higher income than for making charcoal (Srivastava, 1984b).

Forests occupy only 5.1% of the land area in the Punjab; even that is a poor forest. Growing of *Eucalyptus hybrid* on farm boundaries was started in the Punjab in 1967. During the past few years, tree planting on the farms has progressed from line planting on farm boundaries, irrigation channel sides and roads to woodlot planting with intercropping. In 1981-1982 the Punjab farmers used more tree seedlings than used in the Government plantation programmes (Dogra, 1984). In the Punjab, 90% of the cultivated land is irrigated, hence most of the plantations are under irrigated conditions. Even in the remaining 10% cultivated area, the farmers now plant trees. The farmers have responded to the market forces. In 1977, the price of *Eucalyptus* was Rs. 80/m^3 standing; this was reported to be Rs. 300 in 1984. Subsequently, there has been a fall in prices due to excessive plantation and supplies.

In the Punjab, *Eucalyptus* woodlots are raised on the sides of irrigation channels (having a cross-section of 20 × 20 cm to 30 × 30 cm) spaced at 3 to 5 m apart; seedlings are planted at 1.8 m spacing in pits of size 30 × 30 cm cross-section; in case agricultural crops are to be sown a wider spacing of 4 to 5 m between the rows is preferred to facilitate the cultivation of crops by tractor-drawn implements. This is currently restricted to good quality lands. Wheat is by far the main crop; other crops taken are rape seed (*taramira, toria, sarson*) and Bengal gram. Weeding is the major cultural operation done by the farmers. In the case of irrigated *Eucalyptus* plantation without agricultural crops, three to four irrigations are given during the first year and one to two irrigations during the second year; in case agricultural crops are grown, they receive irrigation as per their needs. In the case of unirrigated *Eucalyptus* plantation on undulating land, hand-watering is done twice during the first year, and once during the second year. In the case of irrigated plantation, the use of fertilisers (50 g urea per plant, about six weeks after plantation) is being adopted by the farmers. In the second year 100 g of urea is applied. *Eucalyptus* plantations on agricultural farms have proved to be highly economical (Dogra, 1984) giving an internal rate of return of 35 to 38% without intercropping and 85% with intercropping (Table 11). It is concluded that on good agricultural land where irrigation is available, intercropping must be practised since opportunity cost of this land is high (Dogra, 1984).

Farmers in Gujarat have grown woodlots of *Eucalyptus* in irrigated lands where normally cotton was grown. However, by adjusting the row spacing so as to be able to cultivate with tractors, and by reducing the spacing within the row, they have succeeded in taking an intercrop of cotton for two to three years, and then allowing the trees to grow for two to four more years, to be sold as poles, and then taking two to three coppices. In a case study (Gupta 1979), it was observed that a farmer normally was getting a maximum net income of Rs. 1,250 per hectare from agricultural

Table 11: Economics of *Eucalyptus* plantations on agricultural farms in the Punjab

Land site type	I	II	III	IV
Description of site	Undulating unirrigated land	Fair quality irrigated land (no intercropping)	Average quality irrigated land	Good quality irrigated land (intercropped)
Annual land rent (Rs./ha)	400	600	1000	1500–2500
N.P.V. (Rs./ha) at 6%	26802	31794	42732	47811
15%	8050	9714	13012	17458
I.R.R. %	35	38	37	85

Source: Dogra, 1984

crops like cotton, sugar cane, paddy, wheat and groundnut. As against this, when the farmers grew 64 ha of irrigated *Eucalyptus*, and they took cotton as an intercrop for one year, and then managed the main crop of *Eucalyptus* as well as three coppices, and when the streams of net annual returns were discounted at 11%, the annuity values per hectare were Rs. 6625, Rs. 11,400, Rs. 11,400 and Rs. 11,425 for the initial and the three coppice crops, respectively. The net returns from the coppice crops were more, as should be expected, since there are no initial expenses to establish a plantation. A coppice crop inherits a well-established root, and it matures earlier than the initial crop. In addition, in the above case *Eucalyptus* seeds could be collected twice for each coppice crop, as compared to once for the initial crop.

The price boom in *Eucalyptus* was evident up to about 1984. However, by 1986 the prices had crashed; for example in the Punjab a seven to eight years old tree projected to be saleable at Rs. 150, would not sell even at Rs. 18 in 1988. In Gujarat, the price of a pole of 10-12 cm diameter fell from Rs. 60 in 1986 to Rs. 23 in 1988 (Saxena, 1991). In Haryana, a six-year-old tree could not be sold even at Rs. 10, as against the projected sale price of Rs. 100 (Athreya, 1989). The collapse of the market in the case of farm-grown trees led to a steep decline since 1988 in the planting of *Eucalyptus* in particular, and trees in general, on farmlands. A number of reasons related to over-production and/or market failure are offered for the collapse of the prices (Saxena, 1991). Perhaps one, or more, or all reasons were operative at one or the other time. Without getting into the details of these reasons, it is interesting to note that this traditional practice expanded in response to actual and perceived shortage of trees and their products, but when there was need to sell the products, the market forces, among other things, came into action. Saxena (1991) concludes that the farmers' enterprise seems to have been thwarted by market constraints, and rigid laws. However, one could say that a lesson has been learnt, and the economic forces are bringing in the appropriate corrections.

2.3.3 Other Tree Species

Farmers in Assam grow woodlots of bamboo along with paddy fields.

There are some tree species which are grown in restricted/ specific areas/conditions. Some of the examples are growing of *Bambusa arundinacea* in depressed locations and waterlogged areas in Andhra Pradesh. These are harvested from the sixth year onwards and fetch a very high price. In the dry district of Anantpur, Andhra Pradesh, farmers do budding of *ber* (*Zizyphus* spp.) which grows wild in the fields. The *ber* plants give high returns as compared to the uncertain poor agricultural crop yields. In Cuddapah, a drought-prone district of Andhra Pradesh, farmers grow Redsanders (*Pterocarpus santalinus*) to get poles as shafts for carts and ploughs. The tree is harvested in 12 to 15 years and coppices vigorously (Reddy, 1981).

Farmers of Thanjavur district in Tamil Nadu grow bamboo (*Bambusa*) on small pieces of land that are elevated from the surrounding fields. A few cultivars are identified—one has dense thorns all over the culms, another has less thorns, and still another is devoid of thorns up to a height of approximately 3 m (Shanmugasundaram, 1984).

Woodlots of *Dalbergia sissoo* and bamboo are common in Purnea district of Bihar (Tripathi, 1984).

Farmers raise woodlots both for biophysical and economic considerations.

The number of tree species grown or used by farmers is much more extensive than mentioned here. In the absence of detailed information or description, the next best thing is to be at least aware of them. These species are listed in Appendix 2.

CHAPTER 3

PLANTATION CROPS UNDER SHADE OF TREES

There are a number of plantation crops which require shade for their optimum growth. Among them are tea, coffee, cardamom (small and large), pepper and cacao. Three types of practices can be recognised: (1) shade trees may be planted for growing a plantation crop (for example, tea and coffee); (2) shade may be provided to a plantation crop (for example cacao) by a commercial tree crop (such as coconut and arecanut); and (3) shade may be provided by trees in a natural forest to a commercial crop (for example, small and large cardamom). All the three practices comprise perennial shade trees and perennial trees/shrubs/vines and are good examples of mixed cropping in an agrosilviculture system. Coconut with cacao, and cardamom under forest trees also qualify as farm-based silvoagriculture and forest-based silvoagriculture practices, respectively.

3.1 TEA

India raised tea pLantations on 420,000 ha of land and produced 715,000 tonnes of tea, in 1989-90 (MOA, 1992b). Tea (*Camellia sinensis*) is grown in large commercial plantations of 200 to 400 ha in Assam, West Bengal and Tamil Nadu. These plantations are registered as "Tea Estates" and locally called "Tea Gardens". There are a few small holder tea plantations in India. Tea estates not only grow their own tea, but they also process and manufacture it. They require large capital and hence they are owned and managed by industrial houses. It was recognised in, as early as 1863 that Assam plant *Camellia sinensis* var. *assamica* does not appear to thrive in inferior soils when exposed to full influence of the sun, and partial shade of tea bushes was recommended (Masters, 1863). Again, in as early as 1885, the definite benefits of the *sau* tree of Assam (*Albizia stipulata* syn *A. chinensis*) were published (Buckingham, 1885), and in 1903 "slight shade" for tea was recommended for the lower part of the Brahmaputra valley, except where mosquito blight was abundant (Watt and Mann, 1903). Currently, apart from *Albizia stipulata, Albizia odoratissima, A. lebbek, A. lucida, A. procera, Acacia lenticularis, Derris robusta* and *Dalbergia sericea* are recommended to be grown in mixed stands of shade trees in northeast

India (Dutta, 1977). In South India silver oak (*Grevillea robusta*), *Acacia* spp., *Albizia* spp., *dadap* (*Erithrina lithosperma*) and *Indigofera teysmanii* are used as shade trees.

In actual practice when the land is cleared for planting tea, all woody and other vegetation is removed and the ground is ploughed deep. Tea seedlings or clonal plants about 12 to 18 months old are planted in a rectangular or square pattern to give a population of 10,000 to 12,000 bushes per hectare. Shade tree seedlings are also planted in the rows of tea at a distance of 6 to 8 m in a square pattern depending on the spacing of the rows of tea. After several years (usually seven to 20) alternate shade trees in each line are removed leaving a final spacing of 12 to 18 m. The shade trees remain until the tea is finally uprooted which may be 50 to 70 years after planting and occasionally even longer (Hadfield, 1974). It was a common practice to plant rows of leguminous shrubs between each two rows of tea for the first two years in the field to provide shade for the young tea plants and use the loppings from these shrubs as a mulch. This is still practised but not as widely as in the past since the species used became very susceptible to pests and diseases which then spread to the young tea.

Shade could be provided by a single or a mixture of tree species. It is preferable to plant shade trees in rows to facilitate the movement of labour. In the case of replanting tea, shade trees could be planted during the land rehabilitation period. In any case, they must be planted at the tea-planting time. In tea gardens in Assam, low depressed areas are given to labour to grow paddy. The subject of tree shade, interaction between the shade of different tree species and tea, use of fertilisers, and the yield of tea, has been extensively experimented on in India (Cooper, 1939; Wight, 1959 a,b,c; Venkataramani, 1973; Hadfield, 1974; Gogoi,1976).

Interactions

Although shade trees are planted in tea, their usefulness or otherwise has always been a matter of debate.

A combination of observations, experience and experimentation has established two facts: (a) generally speaking, unshaded tea bushes give a lower yield than the shaded bushes. For example, the shade of *Albizia stipulata* tree increased the yield of "made" tea by 96% and 33% with and without fertiliser application, respectively; and (b) over-shaded tea bushes gave a lower yield than medium- or lightly-shaded bushes (Wight, 1959a). It has been a practice, therefore, to use shade trees which cast sparse or light shade over tea bushes.

Albizia stipulata was the most suitable tree for this purpose. With the decline in the use of *Albizia stipulata* due to diseases such as canker and red rust, 38 tree species belonging to *Albizia, Acacia, Adenanthera, Caesalpinia, Cassia, Chukrasia, Dalbergia, Delonix, Derris, Enterolobium, Erythrina, Gliricidia, Grevillea, Indigofera, Leucaena, Milletia, Parkia, Peltophorum* and *Piptadenia* genera have been evaluated for northeast Indian tea plantations (Dutta, 1977), of which eight shade-tree species, as mentioned in the first para above, are recommended. Many factors other than the amount of light passing through the canopy must be taken into account when selecting a shade tree: for example, time and duration of leaflessness, pest and disease susceptibility, proneness to storm damage, drought resistance, and the possibility of leaves contaminating the tea by falling into the baskets in which tea shoots are carried. Thorns on caducous branches can cause serious injury to the barefooted tea pickers, and trees otherwise potentially suitable would be rejected on this account. An important part of the investigations is to find trees or a mix of trees which give sparse and even shade. Trees vary considerably in the quality of shade (even or uneven, sparse or dense), the range of light transmitted (16 to 70% of full sunlight), mean light intensity (28 to 64%), diameter of the crown (6.6 to 18 m), number of bushes covered per tree (44 to 160 in 1.5 × 0.75 m hedge planting) (Hadfield, year not known). It is obvious that it is a complex process to determine experimentally a tree species or a mix of tree species; further a single tree species or a mix of tree species will require planting at different spacings. It is now recognised that an ideal shade tree for tea should have a single layer of leaf canopy, with evenly spaced small leaflets which will allow an adequate proportion of light in the form of 'sunflecks' to penetrate the canopy and at the same time cut off as much as 50% of the total sunlight on a bright summer day. The suitable spacing of shade trees is about 10 to 12 m.

Since not only tea, but many other plantation crops such as cacao, coffee, pepper and cardamom are grown under the shade of trees, some description of the shade pattern in tea plantations may be of value in understanding the complexity of the shade and crop interactions.

In northeast India the harvesting period for tea is from April until early December, after which the bushes are pruned and much of the foliage is thus removed. Most of the commonly used shade trees are deciduous and may be without foliage for one to four months during the dry season, but in general the tea bushes and the shade trees come into leaf at about the same time, the middle of February. Foliation of the shade trees is rapid and by the middle of May, they are usually fully foliated. From the time of full foliation the light on the surface of the tea bushes consists of a pattern of areas of shadow, the size of which is determined by the canopy of shade trees and the solar altitude (Hadfield, 1974).

The simplest possible case is found on a completely clear day when the sun is directly overhead and the shaded area is equal to the area of the canopy. But even here the situation is complex and not as simple as it sounds. It will suffice to note that the unshaded areas are illuminated by light of fairly well-known spectral composition and easily measured intensity while the shaded areas receive light of different classes. Part of this 'shade light' would have been depleted of some of its spectral energy by the foliage of the shade tree, the amount of depletion being mainly dependent on the density of canopy. Some light would have been reflected from the leaves and trunks of the trees, and some would consist of sunflecks having the spectral composition of direct sunlight but of varying intensity depending on the size of the gaps in the canopy. The number and size of 'sunflecks', and their contribution to the total 'shade light' would again depend on the foliar characteristics of the shade tree and would obviously differ between shade tree species. On overcast days when there are no visible shadows, and on clear days when the sun nears the horizon, different situations arise. It may also be noted that the spectral relations of the tea bush-shade tree community are very much simpler than those of a multistoried forest or a home garden. The closely-spaced tea bushes are pruned to a horizontal surface to form a complete layer of foliage. In well-managed tea estates the shade stand consists of trees of the same age of a single or mixed species, planted in a regular pattern. Hadfield (1974) has reported the spectral analysis of light and light-shade pattern in a tea plantation (planted in 1920; 1.5 × 1.5 m, 1 m high surface) having a pure stand of *Albizia odoratissima* planted in 1950 at 7.5 × 7.5 m spacing, mean height 12 m, mean diameter of trunk at 1.3 m height, 0.6 m, and average crown diameter 6.5 m.

As is to be expected the total light received by tea bushes during different parts of the day, during different seasons, and by rows situated in different directions is quite variable. For example, in the simplest possible case of a clear day in June, when the shade trees are fully foliated and the sun is almost directly overhead at 1200 hours, the mean light intensity over the tea bushes was 60% of that in the open, and the amount received by the individual rows varied from 35 to 94% in both the north-south and east-west rows; by 1400 hours, the mean sunlight received by the tea bushes was 57% of that in the open and the amount received by the individual rows varied from 31 to 78%; these values further decreased to 43%, and 21 to 62%, respectively, by 1500 hours (Hadfield, 1974).

The hours of bright sunshine are markedly reduced during the monsoon and this has a great effect on the shade pattern. The term 'overcast' covers a wide range of conditions, from a light haze over the sun's disc to the virtual darkness of an impending monsoon storm.

The management of tea bushes also contributes to the interaction. For fully-efficient tea production it is necessary for each leaf to produce in excess of its own requirements and the remainder is contributed to the bush reserves. To achieve this, each leaf must receive a certain amount of light, and the light value where there is no gain or loss to the leaf is called the 'compensation point'. Below this value the leaf may be considered to be a passenger to the extent that it receives foodstuffs from leaves in a more advantageous position in the bush. Modern planting methods are based on rapid ground cover using more plants per hectare than was the previous practice and hence more leaves are heavily shaded by those of the adjacent bushes than was formerly. For example, in tea bushes planted at 1.2 × 0.6 m, the lower leaves receive only about 30% of light received by the leaves on bushes planted at 1.5 × 1.0 m. When one considers that a fully mature bush may possess up to 10,000 leaves, the proportion of overshaded leaves becomes considerably high, and once such a tea bush is shaded with densely canopied trees, the proportion of overshaded leaves rises still further, until an appreciable percentage of tea leaves are virtually parasitic.

This fact is probably responsible for the disappointing yields of 1,400-1,600 kg tea/ha obtained from most of the young, hedge-planted tea as against the economic expected yield of 2,300-2,700 kg tea/ha (Hadfield, year not known).

Further complications are introduced by using tea bushes or clones with varying leaf characters. For example, at one extreme a bush may have leaves at right angles to the stem, and at the other end it may have fairly small leaves with an upright habit (Hadfield, year not known).

Soil Moisture and Other Edaphic Interactions

It is reported that soil moisture is more in the shaded than in the unshaded areas. For example, soil moisture was either slightly above or well below the wilting coefficient during the dry months of December to April, in South India (Figure 7). In the shaded area the soil moisture was above the wilting coefficient for most of the time. It is argued that during the dry periods, the shade trees may compete with the tea bush for soil moisture. However, this has not been the case. It would appear that if shade is regulated and the trees do not carry a very heavy canopy during the dry weather, and also if the trees have a deeply penetrating root system, they need not necessarily deplete the soil moisture in the top 46 cm of the soil (Venkataramani, 1973). It is further observed that even when a shallow-rooted shade tree like *Erythrina lithosperma* was used, the soil below it not only contained more organic matter (3.1 and 2.1%, respectively at

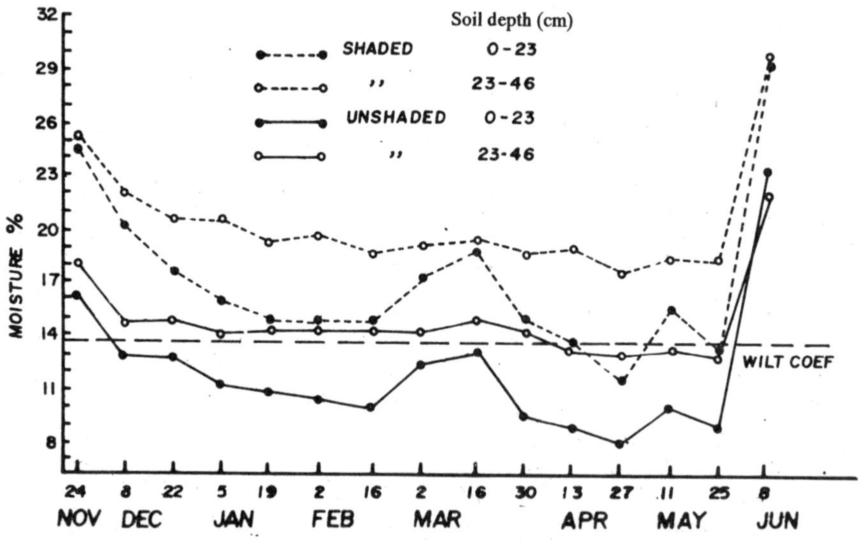

Fig. 7: Soil moisture as influenced by shade trees in tea plantations (Source: Venkataramani, 1973)

0-23 and 23-46 cm soil depth), but also had higher water-holding capacity (56.3 and 54.0%, respectively) as compared to soil without shade (values for soil organic matter being 2.0 and 1.5%, and for water-holding capacity 48.8 and 46.8% at 0-23 and 23-46 cm soil depth, respectively) (Venkataramani, 1973). It is reported from Assam, that yield variations under different shade trees were closely associated with soil moisture at 2.5-45 cm soil depth estimated at the end of March in a drought year (1963) with a strong positive correlation (R = 0.9250). The data available indicate that a properly managed stand of mature, deep-rooting shade trees is unlikely to compete with the tea bushes for soil moisture. Young shade trees, however, are reported to compete with tea bushes for soil moisture (Gagoi,1976).

The roots of shade trees provide better drainage of soil water during the rainy season, leading to better aeration. It is reported that tea grown under the shade of *Albizia stipulata* had a higher proportion of feeding roots in the growing season (April-October) in Assam than tea grown without shade (Barua and Dutta,1961).

Soil temperature is reported to be lower in a shaded area (Gogoi, 1976; Venkataramani, 1973) than in an unshaded area. For example, Venkataramani (1973) observed that soil temperature at 7.5 cm soil depth was 15.9, 18.9 and 19.4° C at 0900, 1200 and 1500 hours, respectively, under the shade, as compared to 17.4, 22.7 and 23.1° C at 0900, 1200 and 1500 hours, respectively, under the unshaded tea bushes. Relatively high soil temperature would reduce the soil moisture and may conceivably contribute to lower soil organic matter, adversely affect the growth of feeding tea roots—(which occur in the top 46 cm of the soil profile), and interfere with the nutrition of the plant (Venkataramani, 1973).

Atmospheric humidity is observed to be much higher in the shade than in no shade. Venkataramani (1973) reported that relative humidity under the shade varied from 62 to 78% during the day, as compared to 39 to 58% without shade. It is surmised that dry desiccating winds in the dry months like January, would adversely affect the unshaded tea. It is well known that in the drought-prone Cachar hills in Assam, the tea bushes under the shade suffer less from drought damage (Heath, 1956). Venkataramani (1973) reported that tea bushes without shade suffered from dry-weather chlorosis and had a lower chlorophyll content than those under shade. It was further observed, that during the driest and hottest part of the year, the stomata were open for a considerable period of time in the shaded leaf than in the leaf fully exposed to the sun. In fact, in full sunlight the stomata remained practically closed from mid-day to 1500 hours. The leaf temperature was also higher by 4-5° C in the case of leaves exposed fully to the sun, while those in the shade had the same temperature as the air. This is now further elucidated by the fact that shade trees reduce infrared radiation reaching the tea leaf surface and thus help in lowering the temperature of the leaf.

It is observed that the net photosynthesis declined sharply at leaf temperatures of 35° C and above, and between 39 and 42° C leaf temperature there was no net photosynthesis. Leaves of tea in full sunshine may reach temperatures of 40-45° C (Gogoi, 1976).

Shade-fertiliser Interaction

Leguminous shade trees benefit tea on two accounts: through the shade, and through the nitrogen they provide. It has been shown that the yield increment due to the shade of the *sau* tree (*Albizia stipulata*), is equivalent to that obtained from about 90 kg N/ha (Wight, 1959a). It is concluded that a satisfactory tea crop yield can be obtained by the use of shade trees alone. If shade trees are not used then larger quantities of nitrogen fertiliser are necessary. Without shade trees, the yield increase is approximately linear up to 135 kg N/ha. Not only is the return per kilogramme of nitrogen less under shade trees, but deviations from the linear response can be expected at a lower dose of ammonium sulphate (Wight, 1959a). Apparently the shade-fertiliser interactions are determined by more basic interrelationships as reported by Clements (1928) who observed that similar states of growth of field pea in controlled laboratory experiments could be attained under different illumination conditions only when the relative proportion of nutrients were changed—these were calcium nitrate, magnesium sulphate and dihydrogen potassium phosphate. There was no implication that these nutrients are the only ones of importance. But the principle that comparable growth under different illumination conditions may necessiate an altered balance of nutrients was demonstrated. One could not suppose this principle to apply only to the plant used by Clements (1928).

Tea is a mycorrhizal plant; it is reported that light intensity is a determining factor in the formation of *Pinus* and *Picea* mycorrhizae (Bjorkman, 1942). Thus it is suggested that certain nutrients become more available to tea plant through its mycorrhizae and the fungal symbiont, as the light intensity is reduced.

Soil Fertility

It is customary in northeastern India to prune the tea bushes in the cold season, usually in December and January, when the bushes are more or less dormant. The weight of the prunings is equal to the weight of the tea leaves and shoots removed for manufacturing the tea. The prunings lie on the ground and form a mull. In the northeast Indian conditions the mull breaks down rapidly, in fact, much of it ceases to exist as mull by the end of the monsoon, and becomes incorporated as organic matter in the soil. It is reported that when the prunings of one season were removed there was a drastic reduction in the yield of tea. It is further reported that the recovery of yield potential under this situation was better under the shade of the *Albizia stipulata* tree than under a bamboo screen shade (Wight, 1959a). This is explained by the fact that many leguminous trees, including *Albizia stipulata* maintain a notably simple branch system. There is an annual succession of caducous stems which fall to the ground and these stems are of more consequence than the fallen leaves and could account for much of the benefit of *Albizia stipulata*.

A shade tree may add as much as 2,500 to 5,000 kg leaf, twigs and pods and 63-126 kg N, 18-36 kg P_2O_5, 22-44 kg K_2O, 32-64 kg CaO and 16-32 kg MgO/ha/annum to the soil (Gogoi, 1976). These nutrients are utilised by the tea crop. No information is available as yet on the nitrogen fixed and made available to the tea bushes by leguminous shade trees. It is surmised that in the presence of large quantities of nitrogen fertilisers applied to tea, the nitrogen fixed by the shade trees may be considerably reduced (Gogoi, 1976).

One of the important effects of shade trees is the favourable partition of growth between different plant organs. The ratios of pluckings to prunings, pluckings to total growth, and shoot to root are also the highest under shade trees (Gogoi, 1976). Shading, therefore, diverts a relatively larger fraction of the assimilates towards production of plucked shoots and thereby increases yield.

Diseases and Pests

Shade trees influence the incidence of diseases and pests, in some cases favourably and in others unfavourably. The absence of shade trees is reported to reduce the incidence of blister blight. However, the interaction is not simple. With the advent of blister blight disease in South India in 1946, regulation of shade was recommended as one of the measures for controlling the disease (Venkataramani, 1973). In Indonesia, where rainfall occurs all the year round, it has been suggested (de Weille, 1959) that shade may be removed altogether to achieve blister blight control, since shade is favourable to blister blight and that sunlight has a deleterious effect on spore germination. It is observed that in South India, there are only six wet months, four dry and two transitional; during the wet months blister blight remains active with or without shade. The only effective method to control it is by the use of fungicides. In view of this, provision of shade to tea is considered essential. Further, the infestation of scarlet mite, red spider mite and other mites and thrips becomes serious in the absence of shade (Venkataramani, 1973; Das, 1961).

Quality and Valuation of Tea

It is observed that the reduction of light to about 50% of full sunlight either by using a mechanical bamboo screen or with tree shade, did not produce any adverse effect on the quality of tea. If anything, the overall character of the tea slightly improved at the lower light intensity (Barua, 1961). It is also reported that there is a slight decrease in the value of tea when yields increase due to fertiliser application or due to shade; however, the total monetary return is much higher in shaded plots than in the full sunlight in all the manurial treatments (Gogoi, 1976).

The above interactions have been examined at length, to indicate that when, not only tea, but other crops are grown under shade, the basic relationships are likely to be the same and may be applicable in varying degrees, to all of them.

3.2 COFFEE

Coffee (*Coffea* spp.) is cultivated commercially in the hilly tracts of the western and eastern ghats in the southern states of Karnataka, Tamil Nadu, Kerala and Andhra Pradesh. *Arabica* (*Coffea arabica*) and *robusta* (*Coffea canephora*) are the two principal economic species of coffee which are extensively cultivated in India: 90,211 ha (58%) were under *arabica* and 65,365 ha (42%) under *robusta* as reported in 1980. In 1990, 243,000 ha were under coffee and 215,000 tonnes of it were produced (MOA, 1992b). The total number of growers was 54,365, the majority of whom were classed as small growers, each with 10 ha or less under cultivation (Bavappa, 1980).

Edaphic Requirements

The annual rainfall ranges from 1,250 to 3,000 mm received mostly during the southwest monsoon. The plant grows well at temperatures between 12 and 36° C. Coffee grows well in lateritic (alfisols) soils—deep, friable, porous, rich in organic matter, moisture-retentive and slightly acidic—pH 6.0 to 6.5. *Coffea arabica* grows well at an elevation of 900 to 1,200 m having an annual rainfall of 1,600 to 2,500 mm. *Coffea canephora* or *robusta* grows at lower elevations (about 150 m) having an annual rainfall of 1,000 to 2,000 mm.

Both the *arabica* and *robusta* coffees have complex fertiliser schedules and doses depending on the age and expected yield of coffee beans; fertilisers are applied in split doses depending on the growth stage of coffee in the season. *Arabica* is more shade-loving than *robusta* under conditions obtaining in South India. A shade tree should be fast-growing and fast-spreading, to allow a uniform amount of filtered light. Its root system should be deep so that it does not compete with coffee. The trees which serve as alternate hosts to the major pests of coffee must be avoided.

Dadap (*Erythrina lithosperma*) is, by far, the universal temporary shade tree in India. It is always planted with coffee in new clearings. It is a legume that enriches the soil by fixing atmospheric nitrogen. The most popular permanent shade trees belong to the Indian fig and legume families. They are *Ficus glomerata, F. nervosa,* among the figs; and *Albizia stipulata, A. lebbek, A. moluccana, A. sumatrana* among the *Albizias.* Other common shade trees are *Dalbergia latifolia, Artocarpus intergrifolius, Bischofia javanica* and *Grevillea robusta.* Permanent shade trees are generally planted about 12 to 14 m apart. It is desirable to plant a large number at first, and gradually thin them out as they grow and spread out. The trees have to be regulated in such a way that in the course of time, the canopies are about 10 to 14 m above the coffee plants. Shade trees require constant attention by way of pruning and lopping to furnish the required shade (Bavappa, 1980).

Coffee has recently been introduced experimentally in Arunachal Pradesh. The forest plantation trees *hollock* (*Terminalia myriocarpa*) and *hollong* (*Dipterocarpus macrocarpa*) are used as shade trees. The application of fertilisers and weeding of coffee have boosted the growth of forest trees.

Terminalia myriocarpa appears to be a better shade tree, since it provides uniformly distributed "thin" shade, and the adverse "drip effect" of rainfall is also less due to the smaller leaf size. Coffee can be introduced in these forest plantations when they are 10 years old, since the economic age of coffee is 50 years and the tree rotation is of 70 years (Bhattee, 1984).

3.3 CACAO

Unlike tea and coffee, the cultivation of cacao (*Theobroma cacao*) is of very recent origin. India was importing 700 to 1,000 tonnes of cacao beans to meet the requirements of the chocolate and beverage industries (Wood, 1964). The British East India Company is credited with introducing cacao in India in 1795 and 1880 (Bavappa, 1980). From imports of about 1,000 tonnes in early 1970, India exported 800 tonnes cacao in 1981-1982. The present area under cacao is around 29,000 ha with a production of about 5,000 tonnes (CPCRI, 1986b).

Edaphic Requirements

Cacao is believed to have originated in the tropical rain forests of the Amazon Valley in South America. It has very demanding edaphic requirements. It grows well where the daily maximum temperature does not fall below 10° C and the annual mean not below 21° C. It grows within a range of 15-39° C and temperature around 25° C is considered optimum (CPCRI, 1986b). The main production centres of cacao are therefore confined to 10° latitude north and south of the equator. It grows well from sea level to about 500 m above sea level; above this altitude, the low temperatures interfere with its growth and yield. It comes up best up to 300 m elevation. A minimum of 90 to 100 mm of monthly rainfall and an annual rainfall of 1,500 to 2,000 mm is necessary, unless it is supplemented with irrigation. The microclimate in the cacao plantation is of vital importance. The cacao plant needs a cool, moist soil and a cool, damp atmosphere around it. Since the natural environment of cacao is the lower storeys of tropical forests, it requires shade when young and to a lesser extent when mature. Young cacao crop grows best with 50% of full sunlight. Cacao grows on a wide variety of soils, with a pH of 4.5 to 7.0. It requires deep well-drained soils (Bavappa, 1980). It does not come up on sandy soils where coconut flourishes (CPCRI, 1986b).

Distribution

In India, where a monsoon climate exists, it is difficult to have large areas naturally congenial to the growth of cacao. The shade, soil moisture and microclimate existing in the arecanut and coconut gardens in certain areas were observed to satisfy the edaphic requirements of cacao. It is now grown in those areas of South India where rain is received during the major part of the year and there is a very short dry season. These areas are restricted to Nynad, Trivandrum, Quilon and Kottayam districts of Kerala; Kanyakumari and Tirunelveli districts and Palni hills of Tamil

Nadu; and Chickmangalur, South Kanara and North Kanara districts of Karnataka. Kerala accounts for about 86% of the area under cacao (Bavappa, 1980; CPCRI, 1986b).

Crops and Trees

Growing of cacao as a mixed crop was first attempted with coconut in 1963 (Nelliat et al., 1979) and with arecanut in 1964 (Bhat and Bavappa, 1972). Bhat and Leela (1968b) suggested that cacao may do well as a mixed crop with arecanut on the west coast of India. Currently, 80% of the cacao is mixed cropped with coconut and arecanut. The rest of it is grown as an understorey crop in partially cleared forests. Recently, it has been successfully grown in Arunachal Pradesh in the forest plantations of hollong (*Dipterocarpus macrocarpa*). Unlike in South Indian states, it can be grown without irrigation in Arunachal Pradesh since the rainfall is heavy and well distributed (Bhattee, 1984).

Arrangement of Components

Some of the details of growing cacao are given in sections 4.1 and 4.2 dealing with mixed cropping in coconut and arecanut respectively, where the emphasis is on coconut and arecanut plantations. Here the focus is on the cacao plant.

Cacao when planted in forest lands by thinning and regulating shade, is spaced at 2.5 to 3.0 m in between and within the rows.

A mixed plantation of cacao and arecanut, can be raised by adopting the quincunx method of spacing arecanut at 4 × 4 m and the cacao occupying the centre of the square; or by spacing arecanut at 2.7 × 2.7 m and cacao at 2.7 × 5.4 m. Both the arecanut and cacao seedlings need shade during the first two years after planting. Banana and *Sesbania* are found to be good shade plants at this stage. They may be planted between and within the rows. *Sesbania* can be used as green manure. During the subsequent years, arecanut palms would provide the required shade for cacao.

When cacao is raised as a mixed crop with coconut (7.6 × 7.6 m), it is raised in a single hedge system at 2.7 m spacing or in a double hedge system at 2.5 m spacing between the rows and 2.7 m between the cacao plants. A study of the rooting pattern of cacao and arecanut mixed cropped at 4 × 4 m in the quincunx system (Bhat and Bavappa, 1972) revealed, that the roots of cacao and arecanut did not compete with each other for soil space and plant nutrients. The horizontal distance available to the roots of any two cacao or arecanut plants with respect to each other was 2 m and between any two arecanut and cacao plants was 1.41 m. Horizontally, 95% of all roots, 80.1% of fine roots and 99.8% of thick roots of cacao were restricted to only 1 m distance; with respect to arecanut, 92.5% of all roots, 89.8% of fine roots and 97% of thick roots were restricted to 1 m, well within the space available to the roots of each plant (Table 12). Vertically also there were differences in the rooting pattern of these

Table 12: Distribution of roots in mixed cropping of cacao (4 × 4 m) and arecanut (4 × 4 m) in the quincunx system of planting

	Cacao roots, %			Arecanut roots, %		
	All roots	Fine roots	Thick roots	All roots	Fine roots	Thick roots
Distance from tree (cm)*						
10-25	48.4	28.5	55.1	60.0	50.0	75.9
25-50	32.8	26.6	34.9	18.9	22.2	13.8
50-75	9.6	13.4	8.3	8.2	10.5	4.6
75-100	4.1	11.6	1.5	5.4	7.1	2.7
100-150	4.2	16.0	0.2	4.8	6.2	2.3
150-250	0.9	3.9	–	2.7	4.0	0.7
Depth of roots (cm)						
0-10	8.4	24.8	2.9	31.1	32.9	28.2
10-20	63.2	22.2	77.1	25.0	24.1	26.3
20-30	9.6	11.3	9.0	11.8	8.7	16.7
30-40	8.8	11.3	8.0	8.5	8.1	9.2
40-50	4.3	10.0	2.4	7.5	6.6	8.9
50-70	3.3	10.9	0.6	7.6	8.1	6.7
70-100	2.1	8.3	–	6.8	8.8	4.0
100-150	0.3	1.2	–	1.7	2.7	–

Source: Bhat and Bavappa, 1972

*Diameter of the tree, 10 cm, excluded

two crops. However, since they did not compete for horizontal space, they were free to explore the soil depth as per their needs.

Resource Input and Utilisation

An application of 100 g N, 40 g P_2O_5 and 140 g K_2O per cacao tree per year in two equal split doses is recommended. During the first and second years the cacao plant may be given only one-third and two-third dose respectively of the fertiliser. From the third year onwards the full dose of the fertiliser is applied. The first dose of the fertiliser is applied in February-March and the second dose in September-October. The fertiliser is applied uniformly around the base of the trees up to a radius of 75 cm. Cacao plants need continuous supply of moisture for their optimum growth and yield. During summer or drought they will have to be irrigated at weekly intervals. Coconut and arecanut have to be manured and fertilised as per their needs as discussed in sections 4.1 and 4.2, respectively (CPCRI, 1986b; Bavappa, 1980).

Cacao trees need to be pruned to develop a good shape. For this purpose all the fan branches arising from the main stem are nipped off. Thin and dried branches are periodically removed. The height of cacao trees is maintained at about 3 m, at second storey level for better light/shade management and ease in spraying and harvesting operations.

More than 50 pests have been recorded on cacao in India. The more important ones are: mealy bugs (*Planococcus lilacinus* and *Pseudococcus citri*); stem borer (*Zeuzera coffeae*); aphids (*Toxoptera aurantii*); stem girdler (*Glenea* spp.); leaf eating caterpillars, leaf beetle (*Popillia complanata*), rats and squirrels.

Among diseases, black pod disease (*Phytophthora palmivora*), canker (*Phytophthora palmivora*), charcoal pod rot (*Botryodiplodia theobromae*), and pink disease (*Pellicularia salmonicolor*), are of importance. Severe deficiency of zinc is observed in India (CPCRI, 1986b).

Ripe pods after harvesting have to be processed for the collection of beans, which are then fermented and dried.

Yields

Yields of some mixed cropping experiments of cacao with coconut and arecanut are discussed in sections 4.1 and 4.2, respectively. However, with a view to optimise the yields of cacao and arecanut in relation to space, a spacing trial was initiated wherein six spacings were evaluated (Table 13). The spacing relationships for individual crops appear to be a function of population; for example, it seems that with a population of 343 or 514 cacao plants/ha the yield of cacao pods was rather low; however with a population of 657 to 1371 cacao plants a stable yield was obtained. With a low population of cacao plants and high population of arecanut, the highest yield of arecanut (though not significantly different from 686 plants of cacao) was obtained. Overall, it appeared that 686 plants each of cacao and arecanut, each at a spacing of 2.7 × 5.4 m gave the maximum stable yield of both the crops. As reported in arecanut mixed cropping (section 4.2) arecanut at 2.7 × 2.7 m and cacao at 5.4 × 5.4 m as mixed crop is recommended to be the most economical combination.

Table 13: Effect of plant spacing on yield of cacao and arecanut in mixed cropping

Spacing (m)		Number of trees/ha		Mean yield (kg/ha)			
Arecanut	Cacao	Arecanut	Cacao	Arecanut		Cacao pods	
				1981*	1983**	1981	1983
2.7×2.7	2.7×2.7	1371	1371	6921	9250	10131	11390
2.7×5.4	2.7×5.4	686	686	9848	10210	10485	9260
2.7×2.7	5.4×5.4	1371	343	12093	12360	7119	5720
3.9×3.9	3.9×3.9	657	657	5893	6070	10682	8470
3.3×3.3	3.3×3.3	918	918	5866	8210	8549	10610
1.8×5.4	3.6×5.4	1028	514	7664	8310	8008	7260
CD for spacing				3175	3480	2486	2260

Source: Bhat, 1982*; Bhat et al., 1983**

3.4 PEPPER

Black pepper (*Piper nigrum*) is one of the most important and the earliest known spices. It is native to the tropical western ghats of South India. In 1990-1991 in India, the total area under pepper was 173,000 ha and the total production was 49,000 tonnes. Pepper cultivation is primarily in Kerala (98%), Karnataka (1.4%), Tamil Nadu, and to a very limited extent in Pondicherry, Assam and West Bengal (MOA, 1992b). Till the fifties, India's share in the world pepper production used to be about 80%; in the eighties it was about 18.3%. Indian pepper is considered to be of superior quality and consequently it fetches a higher price in the international market (CPCRI, 1986c; Bavappa, 1980).

Pepper is a plant of the humid tropics. It grows well between 20° north and south latitudes, requiring an annual well-distributed rainfall of 1,250-2,000 mm. It can tolerate a temperature range of 10-40° C and can grow from sea level to an altitude of 1,200 metres. It grows well in a wide range of soils with a pH of 4.5 to 6.0. In its natural habitat it thrives well on red laterite (alfisol) soils. Being a vine of humid tropical forests, it is shade-loving as well as it needs a standard to climb. In pure plantations live standards of *Erythrina indica* are preferred; *Garuga pinnata, Spondias mangifera, Gliricidia maculata,* and *Grevillea robusta* are the other common standards. Pepper is cultivated as a subsidiary crop in coffee, coconut, arecanut and orange gardens. It is grown in home gardens with mangoes, jack fruit, coconut, arecanut, and many other tree species. In the latter case as many as 27 other species are reported to serve as standards (for example, *Ailanthus excelsa, Alstonia scholaris, Anacardium occidentale, Artocarpus hirsutus, Bombax malabaricum, Careya arborea, Coffea* spp., *Dalbergia latifolia, Delonix regia, Enterolobium saman, Erythrina indica, Gliricidia sepium, Grevillea robusta, Leucaena leucocephala, Macaranga peltata, Mangifera indica, Moringa oleifera, Pajanelia longifolia, Pterocapus marsupium, Saraca indica, Spondias pinnata, Swietenia mahagoni, Tamarindus indica, Terminalia catappa, Terminalia paniculata, Tectona grandis,* and *Thespesia populnea* (Abdul Salam et al., 1991). It is also grown on dead standards. Like coconut and arecanut, pepper is a small holder crop (CPCRI, 1986c; Bavappa, 1980). When grown on sloping land, the slopes facing south may be avoided and the lower halves of north and north-eastern slopes are preferred for planting, so that the vines are not subjected to the scorching effect of the southern sun during summer.

In case live standards are to be used, primary stem cuttings of *Erythrina indica*, or stem cuttings of *Gliricidia maculata, Garuga pinnata* or seedlings of *Grevillea robusta* are planted with the coming of the first rain in May-June. They are planted on the edge of pits (0.5 m cube) dug for pepper. Two to three rooted cuttings of pepper per standard are planted in June-July, 30 cm away from the base of the standard. At a spacing of 2.5 × 2.5 m there will be about 1,600 standards/ha. The fertiliser recommendation for pepper at present is 100 g N, 40 g P_2O_5 and 140 g K_2O per vine per year. Only one-third is applied during the first year, two-thirds during the second year and the full dose from the third year onwards. The fertiliser is applied in two split doses, one in April-May and the other in August-September. Besides, organic manure in the form of cattle manure or compost at the rate of 10 kg per vine is applied during April-May. Fertiliser is applied at a distance of 30 cm away from and all around the vine. Application of lime at the rate of 500 g per vine during April-May in alternate years is also recommended.

In case it is grown as a mixed crop with coconut, arecanut or any other tree crop, each one of the crops should receive its own dose of fertiliser and manure.

As the vine grows, its shoots have to be tied to the standards as often as required. The young vines need to be protected from the hot sun during the summer months by providing them with shade. Regulation of shade by lopping the branches of the standards is necessary not only for providing optimum light to the vines but also enabling the standards to grow straight.

Pepper is attacked by many pests and diseases. Among the pests, pollu beetle (*Longitarsus nigripennis*); top shoot borer (*Cydia hemidoxa*); thrips (*Liothrips karyni*); scales (*Lepidosaphes piperis* and *Lecanium marsupiale*); and mealy bugs (*Pseudococcus virgatis*) are important. Spraying of dimethoate 0.05% is effective in controlling the pest infestation. Important diseases are quick wilt or foot rot (*Phytophthora palmivora*), slow wilt, and pollu disease (*Colletotrichum gloeosporioides*) (CPCRI, 1986c).

Over 75 cultivars of pepper are grown in India. Each tract has its own selection of varieties. They differ mainly in the time taken to mature, the length of the spike and the size of the berries. A new hybrid variety Panniyar-I has been developed. It is a high yielder, giving three to four times more yield than the local varieties and is also an early bearer. It can also grow in a wide variety of agroclimatic conditions in the pepper-growing tracts (Bavappa, 1980). However, this hybrid has a tendency to increased vegetative growth and corresponding reduction in yield under extensive shade and higher doses of nitrogen (CPCRI, 1986c). The rapid technique of multiplication of rooted cuttings developed in Sri Lanka, is now increasingly used in India.

Pepper starts bearing in the third year after planting and investment in a monocrop plantation can be fully recovered in the sixth year after planting. Thereafter, a farmer can expect a high return of Rs. 20,000/ha/annum (CPCRI, 1986c). The produce is black pepper. White pepper is prepared by processing either black pepper berries or freshly harvested green berries. The latter yield about 45% white pepper. One hectare of monocrop yields on an average 600-800 kg black pepper per hectare. The yield starts declining about 25 years after planting, although under favourable conditions the vines may live up to 40 to 60 years (Bavappa, 1980). Experimental results of mixed cropping of pepper with coconut and arecanut are discussed in sections 4.1 and 4.2.

3.5 CARDAMOM

Commercial cardamom is of two types, small cardamom (*Elettaria cardamomum*) and large cardamom (*Ammomum subulatum* and *Ammomum aromaticum*). The small cardamom is cultivated mostly in Kerala, Karnataka and Tamil Nadu; large cardamom in the north-eastern Himalaya region (Sikkim, the Darjeeling District of West Bengal, and Arunachal Pradesh). Cardamom, especially the small one, is the most sought-after spice. Cardamom is used for flavouring various preparations of food, confectionary, sweets, beverages and liquors. It is also used for medicinal purposes, both in the allopathic and *ayurvedic* systems. In 1990-1991 the total area under cardamom in India was 117,000 ha and the production was 9,000 tonnes (MOA, 1992b).

Small cardamom cultivation is more organised. As per the Cardamom Board (1986) the total area under cardamom was about 94,000 ha, of which 60% was grown in Kerala, 30% in Karnataka and 10% in Tamil Nadu. It is a small-holding crop. There were a total of 33,143 holdings of which 31,508 were small growers (less than 4 ha). Roughly about 50% of the land is permanently owned by the growers, and the rest is leased by the state government to the growers (ICRI, 1987).

Two varieties of small cardamom are identified: *Elettaria caradamomum* variety major (made up of wild indigenous types) and *E. cardamomum* variety minor comprising cultivars like, "Mysore", "Malabar" and "Vazhukka". These types are grown in different tracts and are mostly identified on the nature of panicles, size of plant and other morphological characters. A mature cardamom plant may be 2-4 m in height. Cardamom generally grows well in forest loamy soils; these are generally acidic in nature, with the pH value ranging from 5.0 to 6.5. Adequate drainage is essential. The natural habitat of cardamom is the evergreen forests of the western ghats. It is found to grow within an altitude ranging between 600 and 1,200 m amsl. In most of the cardamom areas, the annual rainfall is from 1,500-4,000 mm and the temperature ranges from 10 to 35° C. It is propagated by seedlings, rhizomes and suckers. The vegetative method is simple, reliable, and facilitates easy multiplication of selected clumps or types. For planting cardamom in fields, the land preparation consists of removing undergrowth and thinning out excess shade trees or branches of shade trees in the areas selected for planting. Pits of 45 × 45 × 30 cm are dug at a spacing of 1.83 × 1.83 m for Malabar type in Karnataka. In Kerala and Tamil Nadu, a spacing of 2 × 2 m to 3 × 3 m may be adopted. In sloping lands, pits are dug on contours. The pits are filled with a mixture of top soil and compost or well-rotted farmyard manure and 100 g of rock phosphate. The general fertiliser dose recommended is 75 kg N, 75 kg P_2O_5 and 150 kg K_2O per hectare. Cardamom responds to irrigation. Depending on the soil moisture and local climatic conditions, it may be irrigated once in 15-20 days during the summer (Cardamom Board, 1986).

In the initial stages, cardamom plants require sparse shade for proper growth. In areas with sparse tree growth, suitable quick-growing shade trees are to be planted. *Mesopsis emini*, commonly known as the African shade tree, is fast-growing and can be grown as a temporary shade tree. Trees like red cedar (*Cedrella toona*), *balangi* or *kurangatti* (*Acrocarpus fraxinifolius*), and jack fruit (*Artocarpus heterophyllus*) are good shade trees. Trees like *Cedrella toona* and *Acrocarpus fraxinifolius* which shed leaves in the monsoon in the midst of evergreen forests, ensure natural shade regulation. In the selection of shade trees in cardamom plantation tall trees with a well-distributed branching habit and small leaves are to be preferred. Temporary shade trees should be removed when permanent shade trees develop proper canopies.

Nursery pests are rhizome weevil (*Prodioctes haematicus*); shoot borer (*Dichocrocis punctiferalis*); cutworm (*Arcilasisa plagiata*); mites, and nematodes. Cardamom plantations attract as many as 13 pests—thrips (*Sciothrips cardamomi*); shoot and capsule borer (*Dichorcrocis punctiferalis*); early capsule borer (*Jamides* spp.); scolytid borer (*Thampuriqides cardamomi*); hairy caterpillars (as many as eight species, the important ones are *Eupterote cardamomi, E. canarica* and *E. lenoderarittata*); shoot fly (*Formosina flovipes*), white flies (*Dialeurodoes cardamomi*), root borer (*Hilarographa caminodes*); soft brown scale (*Saissetia coffeae*), cardamom scale (*Mytilaspis* spp.), lace wing bug (*Stephanitis typicus*), nematodes, and mites (spotted red spider mite and red sheath mite) (Cardamom Board, 1986).

A number of pest management practices are reported for Kerala and Tamil Nadu. However, while applying these treatments one has to be careful about bee management in the plantation, since the cardamom flower is bisexual, adapted to the entomophyllus type of pollination by bees (*Apis cerana indica* and *A. dorsata*). It is recommended that four bee hives per hectare of plantation should be maintained. Spray formulations are preferred to dust formulations; insecticides may be applied in the afternoon rather than in the forenoon since bees forage between 0700 and 1100 hours.

Nursery diseases of cardamom are: nursery leaf spot (*Phylloslicta elettaria*); damping off; rhizome rot or clump rot (*Pythium* spp. and *Rhizoctonia selani*). Diseases of cardamom plantations are: *Katte* or mosaic or marble disease; clump rot or capsule rot disease (*Phytopthora nicotianae*); clump rot or rhizome rot (*Phythium* spp. and *Rhizoctonia* spp.); and Chenthal disease (*Corynebacterium* spp.). A number of minor diseases are also recognised; for example leaf spots (*Sphaceloma cardamomi* and *Cercospora zingiberi*); leaf blotch (*Phaeodactylum alpiniae*); leaf rust (*Phakospora elettariae*); bacterial canker (*Xanthomona* spp.). Bordeaux mixture is the main treatment of all these diseases (Cardamom Board, 1986).

Shade trees are owned by the state government. Growers can use dead, fallen trees for fuelwood. The normal yield of small cardamom is 100 kg/ha. The plants give an economic yield up to the 12th to 15th years. The benefit : cost ratio is not worked out (ICRI, 1987).

Current research needs include the evolving of high-yielding, and drought-, pest- and disease-resistant varieties, effective control of pests and diseases, developing suitable agro-techniques, and improving post-harvest techniques (ICRI, 1987).

Large cardamom needs humid conditions within a temperature range of 10-33° C, well distributed rainfall of 2,000-3,500 mm and environmental conditions prevailing under the shade of forest trees. It does not do well in drought and frost-prone areas. Evergreen forests and well-drained soil—rich in organic matter, are suitable for its good growth. Three varieties are grown in Sikkim—Sawaney (early, bold capsules, tolerant to Furkey disease), Golshai (late, bold capsules and seed), and Ramshai (late, small capsules and seed).

Large cardamom is cultivated under the shade of natural forest as well as on flat or sloping lands by planting shade trees. Its cultivation under natural forest involves clearing of undergrowth and lopping or thinning of shade trees to provide checkered shade. Footpaths are provided inside the plantation for transport facility of inputs and produce. If it is to be planted in open, flat or sloping fields, shade trees are planted immediately after or along with the planting of cardamom. The crop is planted either as rhizomes or large clumps at a spacing of 1.5 × 1.5 m, to 2 × 2 m. The shade trees used are *Alnus nepalensis, Schima wallichii, Cinchona* spp., *Lagerstroemia* spp., *Albizia lebbek, Albizia julibrissin, Castanopsis tribuloides, C. hystrix, C. indica*, among others. Large cardamom takes four to five years to come to production. The yield reported is 250 to 300 kg/ha. Sikkim produced about 3,258 tonnes of large cardamom from about 11,204 ha in 1980. Large cardamom has a few pests (hairy caterpillar, and borer) and virus diseases (Furkey and Chirkey) (Gupta, 1982, 1983; Upadhaya and Ghosh, 1983). *Alnus nepalensis* apart from providing shade, is reported to fix nitrogen and enrich the soil by adding leaf litter, which in turn benefit the large cardamom (Venugopal, 1986). *Alnus nepalensis* also provides fuelwood and timber. In Arunachal Pradesh, large cardamom is grown under the shade of *Terminalia myriocarpa, Bischofia javanica* and *Alnus nepalensis* (Bhattee, 1984).

Very little information is available regarding the interaction of shade, types of shade trees, number of shade trees, fertiliser needs, interaction of shade and fertilisers on the one hand, and performance of cardamom on the other hand.

3.6 OTHER CROPS

Paan or betel vine (*Piper beetla*) prefers shade for its growth. In Orissa, betel vines are trained on live standards of *Sesbania* species. In Arunachal Pradesh, there is an indigenous variety of *paan* which is consumed locally. The leaf is rather tough and pungent. It is grown in Tirap district. The vines are trained on *Dipterocarpus macrocarpa* trees growing in the forest (Bhattee, 1984).

CHAPTER 4

AGRICULTURAL CROPS WITH COMMERCIAL TREES

Two types of silvoagriculture practices are identified. In one case farmers grow agricultural crops with commercial trees—the latter component being more dominant and more valuable. In the second type, the forest dwellers, the farmers and the forest departments, use forests to grow crops, and to extract non-timber wood products. In this case also trees are the dominant component and the more valuable.

Growing of commercial trees is a very important cash-cropping practice. Commercial tree crops take a long time to yield returns, and therefore, the need is felt to have interim returns. These trees are usually widely spaced and planted in rows and lines. Till they grow and cover the entire land, it is possible to raise agricultural crops in the interspaces. As a consequence, raising of agricultural crops with commercial tree crops is a very widely practised system. The most outstanding traditional examples are intercropping with coconut, arecanut, and horticultural trees. Of late crops are being grown with short-rotation (poplar and *Eucalyptus*) and long-rotation commercial forest tree species, like teak.

4.1 INTERCROPPING WITH COCONUT

Distribution

Coconut (*Cocos nucifera*) grows in the states of Kerala, Karnataka, Tamil Nadu, Andhra Pradesh, Orissa, Maharashtra, Assam, West Bengal, Tripura, Goa, and the Union Territories of Daman and Diu, Pondicherry, Lakshadweep Islands and Andaman and Nicobar Islands. It is the temperature which seems to determine the territorial boundaries of coconut with respect to the altitude and latitude. In 1990, India grew coconut in 1,513,900 ha of land and produced 9,700.2 million nuts. The four sourthern states of India, Kerala, Karnataka, Tamil Nadu, and Andhra Pradesh accounted for 1,384,600 ha which were 91.5% of the total area under coconut, and produced 8,817.8 million nuts which represented 91.9% of the total production of coconut. Kerala, alone accounts for about 57% of the total area and 46% production of coconut in India (MOA,1992a).

Growing annuals or perennials in the interspaces of coconut trees is an age-old practice in India. Krishna Marar (1964) estimated that in Kerala 78% of the coconut area was inter/mixed cropped, of which 20% was with tapioca; in Andhra Pradesh 35% of the gardens were inter/mixed cropped with banana, turmeric, citrus, mango; and in Tamil Nadu 60% of the area was inter/mixed cropped with rice, sugar cane, sorghum, tapioca and vegetables.

Edaphic Requirements

Coconut needs temperature that is not only high but, as constant as possible. A mean annual temperature of 27° C with a diurnal variation of 6-7° C is optimal. A temporary steep fall or rise, which occurs occasionally in all places, does not harm, provided such extremes do not last for long periods at a stretch (Nair, 1979; Bavappa, 1980).

Coconut does well if the annual rainfall is 1,500 to 2,250 mm and is well distributed (Menon and Pandalai, 1960). It can withstand higher rainfall (up to 4,000 mm per annum) provided the soils are well drained. It is not suited to areas with long and pronounced drought spells over long periods. For good growth, atmospheric humidity has to be high (80-90%), the monthly mean not going below 60%. Coconut also requires plenty of sunshine. Annual sunshine period of 2,000 hours, and 120 hours in a month are considered as the limits below which light becomes a limiting factor (Fremond et al., 1966; Bavappa, 1980). Wind plays an important role in the dispersal of pollen and fertilisation of flowers of coconut. Fairly windy atmosphere is conducive for its good growth. However, violent winds can uproot it (Nair, 1979).

Even though coconut is widely spread over a very varied pedological domain of the tropical zone, the best coconut soils are characterised by an open texture, good drainage, adequate supply of plant nutrients and moisture, and at least 1 m deep soil without any hard layer. In India, coconut grows in coastal sandy soils, lateritic soils of the uplands between the coast and hill slopes, alluvial soils, red loamy soils, reclaimed soil of marshy lowlands and coral soils (of Lakshadweep Islands). A water table that is too high and remains static for long periods is harmful (Nair, 1979; Bavappa, 1980). Special land areas, such as mangroves, coral soils, reclaimed marshy lands, littoral sands and raised *bunds*, which are suitable, or made suitable by appropriate measures, for the establishment of coconut plantation, are not suited for economic intercropping (Nair, 1979).

Growth Habits of Coconut

The crown shape and growth habits of coconut necessitate its planting at a distance of about 7.5 m apart or allocating 56.25 sq m of land area per tree. This would accommodate about 175 trees per hectare, although in the cultivators' gardens up to 250 trees per hectare may be planted (Nelliat, 1979a).

Coconut is a perennial tree which due to its height, particular shape of the crown, root pattern and wide space of planting, permits growing of annual, biennial and perennial crops under it. The pattern of the crown and the arrangement of leaves permit sufficient sunlight for the growth of intercrops. The extent of sunlight reaching the ground varies with the age of the coconut tree, number of functional leaves on the crown, distance between trees, and geometry of planting. In the juvenile stand, considerable amount of sunlight reaches the ground; the light incident in the interspaces in coconut stands decreases with the increasing age of the plantation, and the minimum amount of light reaches the ground when the coconut trees are about 10 years old (Figure 8). This intense "shade" condition lasts after about eight years and up to 25 years. Thereafter, an increasing amount of slanted and filtered light reaches the ground, and it gradually increases with the age of the coconut tree. As the height of the coconut tree increases, the shaded ground area decreases and the incidence of sunlight on the ground increases. After attaining maturity there is little change in the spread of the coconut crown, although a slight reduction in the size of the frond may occur (Nelliat, 1979a; Nair and Balakrishnan, 1976).

It is observed that under normal management conditions, the roots of an adult bearing coconut tree growing in a medium-textured soil (sandy loam), are concentrated within a radius of about 2 m around the tree. Vertically, the top 30 cm layer of soil is practically devoid of functional roots and about 85% of the roots are found between 30 and 120 cm soil

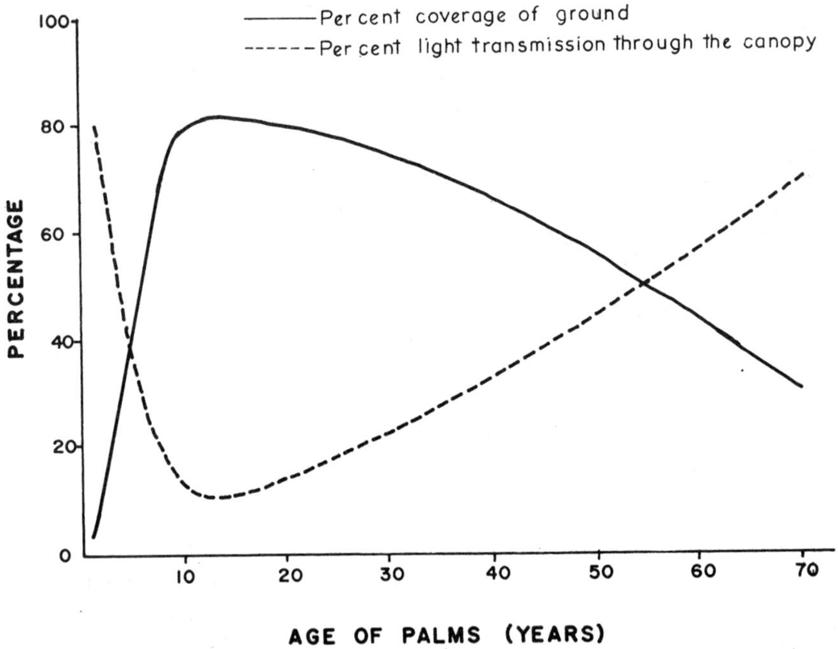

Fig. 8: Interrelationship of age of coconut palms with ground cover and light transmission by canopy (Source: Nelliat, 1979a; Nair and Balakrishanan, 1976)

depth from the surface (Kushwah et al., 1973; Nair, 1979). On surface area basis, the area occupied by a single coconut tree is 56.25 sq m (7.5 × 7.5 m); the area of active root zone is 12.57 sq m (radius of root spread being 2 m). Thus in a pure stand of coconuts, about 75% of land area is not being effectively utilised by the coconut roots.

Crops

Depending on the agroecological conditions and food needs of the farmers, a wide variety of crops are grown such as roots and tubers—cassava/tapioca (*Manihot esculenta*), elephant foot yam (*Amorphophallus campanulatus*), greater yam (*Dioscorea alata*), lesser yam (*D. esculenta*), sweet potato (*Ipomoea batatas*), *taro* (*Calocasia esculenta*), Chinese potato (*Coleus parviflora*); spices—ginger (*Zingiber officinale*), turmeric, (*Curcuma longa*); grain legumes—green gram (*Vigna radiata*), black gram (*Vigna mungo*), pigeon pea (*Cajanus cajan*), horse gram (*Dolichos biflorus*), cowpea (*Vigna catjang*), soybean (*Glycine max*); paddy (*Oryza sativa*); and fruits—banana (*Musa paradisiaca*), pineapple (*Ananas comosus*). In the low rainfall areas of Karnataka and Tamil Nadu, finger millet (*Eleusine coracana*), sorghum (*Sorghum vulgare*) and maize (*Zea mays*) are also grown.

Trees and Other Woody Perennials

There are two distinct varieties of coconut—tall and dwarf. Tall is the economic variety, cultivated extensively for copra, oil and fibre. Dwarf is usually planted for tender nuts and ornamental purposes. Apart from growing the above agricultural crops, the practice of growing tree crops with coconut is also prevalent. Mango (*Mangifera indica*), jack fruit (*Artocarpus heterophyllus*), arecanut (*Areca catechu*), bread fruit (*Artocarpus alitilis*) and sometimes cashew (*Anacardium occidentale*) are grown mixed with coconut.

Animals

Animals/livestock traditionally are not integrated with this practice, except that the farmer may recycle the agricultural waste as fodder for the animals maintained by him.

Arrangement of Components

In time, the life span of the coconut tree could be divided into three distinct phases from the point of view of intercropping. The first phase lasts from planting till the full development of the crown (about eight years). During this period, when the crown size increases gradually, the interspaces could be utilised for intercropping with annuals or other short-duration crops which do not compete with the developing coconut trees for edaphic requirements. The period from about 25 years age is the second phase when the canopy covers about 80% of the ground and the crowns are also low due to the short trunks of the palm. Due to dense shade, there is little or no scope for growing other crops in the interspaces during this period. After 25 years of age there is a gradual increase in the magnitude of the penetration of light to the ground due to the increased trunk height and decrease of the ground coverage by the crown canopy. The period after 25 years of age is the most important phase from the point of view of intensive intercropping.

In terms of space, the practice of growing annual or relatively short-duration crops in the interspaces of perennials is referred to as intercropping; whereas growing other perennials in the interspaces of perennials is called mixed cropping in India. The term "multistoried cropping" is used to refer to multispecies crop combinations involving annuals and perennials with an existing stand of a perennial. In the case of multistoried cropping, the space is utilised in three dimensions to take advantage of the height of the coconut crowns. For example, in a four-crop multistorey cropping of coconut, pepper, cacao/cinnamon and pineapple, the air space is utilised vertically at four levels: the coconut crown occupying the highest level; at the lower level, pepper has its canopy from 2-8 m on the coconut trunk, the canopy of cacao and cinnamon is pruned and maintained at about 3.5 m above the ground level, and pineapple occupies the lowest level near the ground. All these crops are so spaced horizontally, that they benefit from and receive the light which reaches them through coconut crowns (Nelliat, et al., 1974). The concept of three-dimensional utilisation of space is extended not only to the above-ground space, but also to the below-ground space. It was mentioned earlier that in a pure coconut stand about 75% of the area is not effectively utilised by the coconut roots. This space can be utilised by introducing a variety of crops which would also explore the soil for plant nutrients in three dimensions. The exploration of soil for plant nutrients in a multistoried cropping of coconut with cacao/cinnamon, pineapple and pepper is depicted in figures 9a and 9b. It was observed that the roots of the four types of plants did not compete among themselves for space and plant nutrients (Nelliat, et al., 1974, 1979b). The cacao and cinnamon roots had relatively less lateral spread; a three-and-a-half-year-old cacao plant had most of its roots within a radius of 80-100 cm, and the maximum concentration was between 20 and 60 cm laterally (CPCRI, 1974). The fibrous roots of pineapple rarely extended beyond 50 cm laterally or vertically.

Currently, inter- or mixed cropping with coconuts is mostly done in a haphazard manner without giving adequate attention to the edaphic requirements of coconut as well as of other crops—this is much more true of multistoried cropping.

Resources Input and Utilisation

Coconut is essentially a crop of small growers in India. The distribution of coconut holdings in the four southern states (Kerala, Karnataka, Tamil Nadu and Andhra Pradesh) indicates that over 90% of the coconut holdings were either less than 0.2 ha or not more than 1.0 ha (Table 14). Hardly 2% of the holdings had an area of two or more hectares (DCD, 1970). Coconut holdings are looked after with varying degrees of care.

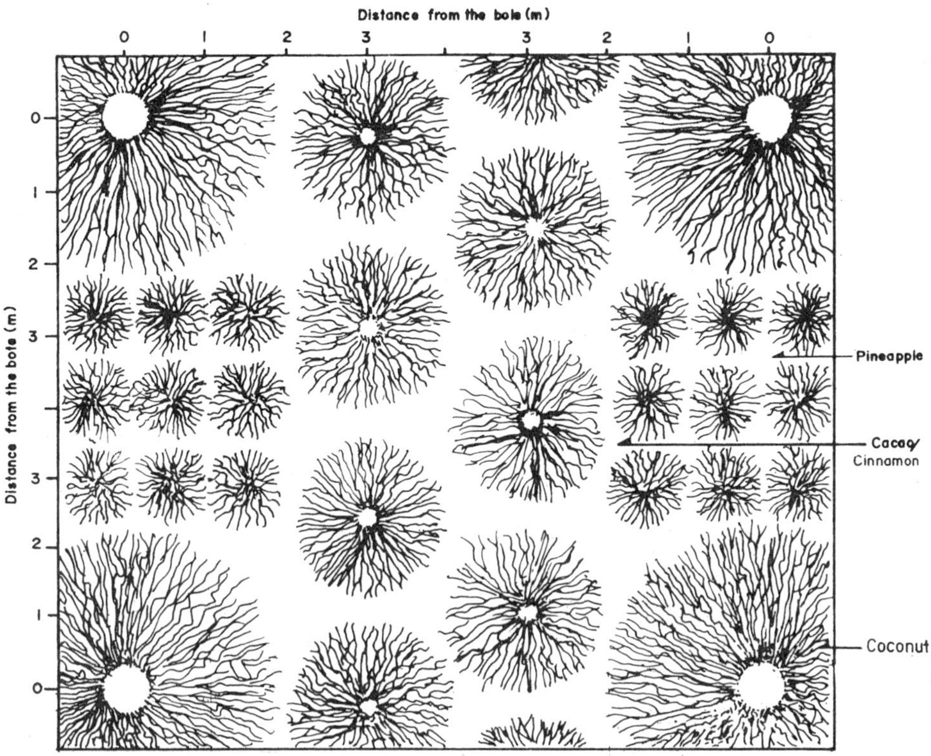

Fig. 9a: Horizontal root distribution in multistoried plantation of coconut, cacao/cinnamon, pepper and pineapple (Nelliat et al., 1974)

Table 14: Distribution of coconut holdings

Area of holding (ha)	Percentage of holding in states			
	Kerala	Karnataka	Tamil Nadu	Andhra Pradesh
Less than 0.2	37.1	61.5	52.5	61.7
0.2–1.0	52.8	33.2	42.9	36.6
1.1–2.0	7.9	3.6	3.6	1.7
More than 2.0	2.2	1.7	1.0	–

Source: DCD, 1970

Coconut lends itself to small-scale production and relatively easy management. The labour required for the crop can be spread fairly evenly over the year and almost all the cultural operations are much easier than those in arable farming. The majority of small holdings command the labour of their families, for whom there is usually no alternative but to work on their land. It is estimated that in Kerala, which accounts for about 57%

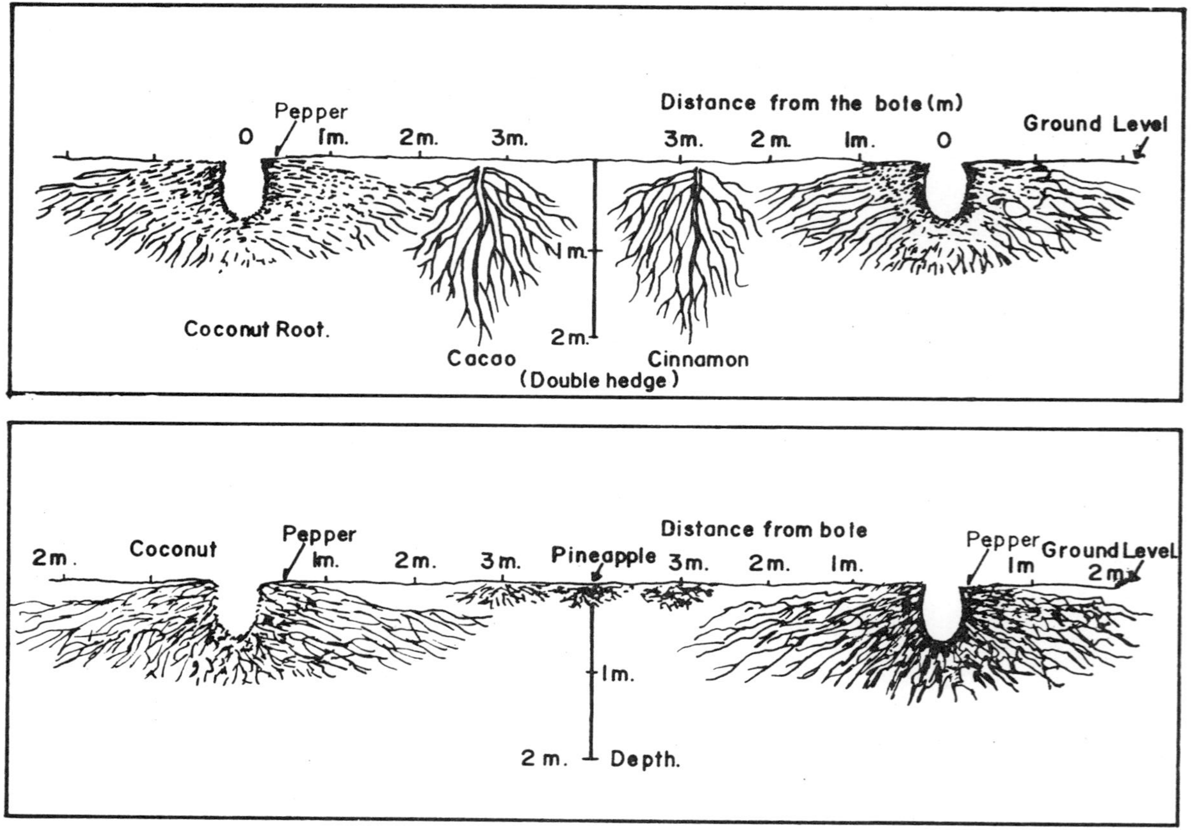

Fig. 9b: Vertical root distribution in multistoried plantation of coconut, cacao/cinnamon, pepper and pineapple (Nelliat et al., 1974)

of the total area under coconut in India, about 70% of its population depends, at least partly, on coconut in one way or another for its livelihood (Nair, 1979). A great number of holdings are owned by part-time farmers, or people engaged in non-farm work who consider it safer to invest in coconut holdings which provide secure, tax-free income. They engage labour to do the farm work.

Coconut plantation is labour intensive. Das (1984) reported that 256 mandays/ha are required during the first year, when the coconut plantation is established; the major one-time job being land clearing, marking and digging pits, and fencing (total 160 mandays), and irrigation (42 mandays). The mandays needed in subsequent years are 115 for the second year; 126 for the third year; 87 per annum for the fourth to the seventh years; 104 per annum for the eighth to the 60th years. These values do not include harvesting of coconuts since it is done by skilled climbers, who are paid higher wages at the rate of the number of nuts harvested. Normally, no fertilisers and manures are added. However, there is experimental evidence to suggest that it is desirable to fertilise coconut as well as intercrops for the best results.

The cost of cultivation of one hectare of coconut, at the 1976 cost level, was Rs. 6,270 in the first year and Rs. 25,235 up to the end of the tenth year. Das (1984) estimated that the total cost of bringing one hectare of coconut plantation to bearing stage was Rs. 35,300, excluding the cost of land estimated to be Rs. 50,000/ha. This is a very high cost of establishment for a small holder. It is reported that the "pay back period" is about 16 years (KSPB, 1973). However, these figures do not apply to replanting (also called underplanting) on existing plantations, which is more commonly practised. In small holdings, replanting is done in a phased manner by underplanting a few coconut seedlings each time, at intervals of one to a few years, and removing the old trees after the underplanted trees have commenced bearing, so that the steady income from a garden is not seriously disturbed (Nair, 1979).

In a well-managed, rainfed, bearing plantation, the major items of expenditure are fertilisers (44%), harvesting and cleaning of crowns (20%), and interculture and miscellaneous operations (31%). Even though plant protection operations cost only 5%, they are the most vital for the success of a plantation (Nair, 1976). For irrigated plantations the expenses are: 30% for irrigation and fertilisers, 13% for harvesting and cleaning the crowns, 22% for intercultural and miscellaneous operations, and 4% for plant protection.

On a rainfed plantation of coconut alone a farmer will make a net profit of Rs. 4,000/ha per annum, and on an irrigated plantation Rs. 6,050 (Table 15). Coconut plantation requires monthly operations for maintenance. The farmer also receives an income in every month except in September and October in the case of rainfed coconut, and in September in the case of irrigated coconut.

Table 15: Month-wise cost of maintenance and net income (Rs.) from one hectare of coconut under different cropping patterns

Cropping pattern	Particulars	Month												
		Jan	Feb	Mar	Apr	May	Jun	Jul	Aug	Sep	Oct	Nov	Dec	Total
Coconut rainfed	Expenses	50	50	70	40	440	30	30	350	750	380	120	8	2400
	Net income	370	540	632	800	332	516	486	124	–330	–14	228	316	4000
Coconut irrigated	Expenses	265	275	285	180	440	30	30	350	750	380	220	295	3500
	Net income	365	613	753	1086	668	78	738	358	–102	178	296	306	6050
Coconut+ cacao	Expenses	295	815	315	618	918	70	50	895	614	400	850	360	6200
	Net income	855	685	1515	1682	1482	1950	970	50	246	660	1030	1360	12500
Multistoried	Expenses	315	895	375	673	1156	100	300	905	694	940	860	345	7520
	Net income	1095	1175	2485	3132	1892	1920	720	40	166	190	1020	1590	15430

Source: Nair, 1976

A coconut plantation is considered an investment, which a family usually builds up with much unpaid labour supplemented by a small amount of cash.

Yields of Intercrops and Mixed Crops

Data on the yield of coconut and other crops growing under it in the farmers fields are not available. Studies on intercropping in coconut were initiated in the thirties at Coconut Research Stations at Kasargod and Pilicode. Research on intercropping has been intensified over the past three decades (Gopalasundaram and Nelliat, 1979). Available experimental data have been reviewed by Nelliat and Bhat (1979).

Tubers as intercrops

In an experiment, rice, tapioca, colocasia and groundnut were grown as intercrops in a 50-year-old coconut plantation over a period of seven years (Table 16). As a result of intercropping there was an increase in the yield of coconut. There was an additional income from coconut as well as the intercrops (Kannan and Nambiar, 1976). Among the intercrops, tapioca gave the highest profit, and colocasia provided the maximum return from coconut; however coconut/rice, coconut/tapioca and coconut/colocasia combinations were comparable in total net profits. It is reported that when elephant foot yam and tapioca were grown as intercrops and fertilised, in a coconut plantation which had passed its prime bearing age, it was necessary to manure both the coconut and the intercrop and give the recommended management practices, as if they are to be grown as sole crops, so as to ensure satisfactory yield and profitability (Varghese et al., 1978a). Manuring of coconut as well as the intercrop, helped in sustained yield of the intercrop and coconut. If only the intercrops were manured, then there was a reduction in the yield of coconut and the yield of the intercrop was also less than that when both the coconut and intercrop were manured.

Table 16: Yield and returns from different intercrops with coconut

Inter-crop	Yield of coconuts (nuts/palm/year)		Yield of intercrop (kg/ha/yr)	Addl. returns from coconut (Rs./ha/yr)	Profit from intercrop (Rs./ha/yr)	Total additional net profit (Rs./ha/yr)
	Pre-treatment (1960-66)	Treatmental period (1968-74)				
Rice	59.9	70.8	850	1164	885	2019
Tapioca	55.9	61.0	15452	544	1503	2047
Colocasia	45.1	58.8	6250	1463	715	2178
Groundnut	58.9	60.5	965	150	455	605

Source: Kannan and Nambiar, 1976

There is experimental evidence to suggest that the growing of tapioca or elephant foot yam, year after year, or even in alternate years, may decrease the yield of coconut (which was 65 years old when the experiment was conducted); however when a five-year rotation of tapioca, elephant foot yam, sweet potato, ginger and turmeric was evaluated, elephant foot yam increased the yield of coconut. The yield of intercrops was also high and sustained when they were rotated. Intercropping coconut with annual crops generated additional employment and profitability (Table 17). All the intercrops had a high output : input ratio, varying from 1.38 to 3.01, except turmeric which had less than 1. Among tubers, elephant foot yam had the highest output : input ratio (due to intercrop) of 3.01 and also the highest net income (including that of coconut), and high employment generation. Among the spices, ginger was better and also the second best among all the annual crops, giving an output : input ratio of 2.06, high income and high employment generation (Varghese et al., 1978a). In another experiment, the maximum profit was reported when elephant foot yam was intercropped with coconut (KAU 1979). Elephant foot yam doubled the employment potential per unit area of land compared to the sole crop of coconut (Gopalasundaram and Nelliat, 1979). It is clear from the above that the farmers have sufficient incentives of economic returns, employment generation, and availability of food that they prefer elephant foot yam and tapioca as food crops in Kerala.

Table 17: Net returns and additional employment generated per year from one hectare of coconut intercropped with tuber crops and spices

Intercrop	Yield of intercrop (kg/ha)	Value of intercrop (Rs./ha)	Net return from inter-crop alone (Rs./ha)	Output: input ratio for inter-crop**	Total net income including coconut (Rs. ha)	Additional employment	
						Man-days	Woman-days
Coconut alone	–	–	8400	–	5500	–	–
Tapioca	11200	4480	2358	2.11	7858	93	12
Elephant foot yarn	12200	9760	6514	3.01	12014	123	8
Sweet potato	7100	2840	785	1.38	6285	56	26
Ginger	7800	9750	5020	2.06	10520	108	28
Turmeric	8100	3248	1808	0.71	3692	108	24
Chinese potato	5600	4480	2445	2.20	7945	95	23
Greater yam	10300	5150	2323	1.82	7823	64	12
Lesser yam	7500	4500	1673	1.59	7173	64	12

Source: Varghese et al., 1978a

*Net return = Value of crop-cost of cultivation

** Output : input ratio = Value of intercrop ÷ cost of cultivation

Rice

Efforts were made to grow rainfed rice as an intercrop since 1934, but the practice was not found to be economical due to low yields from 423 to 749 kg/ha (DAM, 1942). In 1976-78, new rice varieties of 110 days duration were tried when a yield potential of 1,231 to 1,646 kg/ha was achieved (Gopalasundaram and Nelliat, 1979). If one is to go by the results reported by Kannan and Nambiar (1976) quoted above, rice crop apart from being the staple food of the people in Kerala, is also remunerative as an intercrop.

Banana

As was reported earlier (Krishna Marar, 1964), banana was popular as an intercrop in coconut plantations in Kerala, Tamil Nadu and Andhra Pradesh. Wherever, irrigation facilities were available, banana was the most common intercrop in Andhra Pradesh (Anonymous, 1964). Selection of a proper banana variety is reported to be important (Gopalasundaram and Nelliat, 1979; KAU, 1979; Anonymous 1978b). As with any other intercropping system, banana is reported to generate an additional employment potential of 132 mandays and to give a net profit of Rs. 2,905 per hectare (Krishnaji et al., 1976).

Millets

A lot of interest was evinced in the possibilities of growing millets as intercrops in coconut gardens but the yields were very low. For example, *Eleusine coracana* yielded only 236-292 kg grain/ha (DAM, 1932); some information on the yield of *Eleusine coracana, Pennisetum typhoides, Paspalum scrobiculatum, Panicum miliare, Echinochloa frumentacea* has been reported (DAM, 1941, 1942, 1943), indicating that sowing at the beginning of the southwest monsoon was better than late sowing; that among the five millets tested, *Paspalum scrobiculatum* gave the highest grain yield, and that all the millets gave high yield of straw, perhaps due to a long growing season under high rainfall. One lacuna of this experiment was that no yield of coconut was reported to evaluate the whole system. Due to these reasons, and also because of the availability of high yielding and more acceptable tuber crops and other cash crops, the interest in millets as intercrops has diminished.

Pulses

Trials on intercropping coconut with pulses were started as early as in the thirties (DAM, 1932, 1934, 1941, 1942) when horse gram (*Dolichos biflorus*), black gram (*Phaseolus mungo*), green gram (*Phaseolus aureus*), pigeon pea (*Cajanus cajan*), pillpesara (*Phaseolus aconitifolius*) and cowpea (*Vigna catjang*) were evaluated. The grain yields obtained were very poor and variable, 19 to 454 kg/ha. Again in the seventies when new varieties of pulse grains were available, cowpea varieties C152 and C448, and pigeon pea hybrid 3C gave satisfactory yields, and other pulses (green gram, black gram, soybean and horse gram) gave poor yields (Anonymous, 1978a). It has been observed that August to early September sowings, when the heavy rains are over, are conducive to better yield; for example cowpea variety *Kunnamkulam* gave a yield of 467-500 kg/ha (CPCRI, 1977). It would appear that for stable and high pulse grain yields, the suitability of a pulse crop, its appropriate variety, time of sowing etc. are very vital. For example, crops like green gram and horse gram which are more suited to semi-arid conditions, may not be suitable as intercrops with coconut.

Oil Seeds

Among the oil seeds evaluated as intercrops in coconut, castor has proved to be unsuitable (DAM, 1934; 1941) as it was found to be susceptible to shade; sesamum proved to be partially successful giving a yield of 278 kg grain/ha with late June sowing. Sahasarnaman (1964) reported a yield of 600 kg for groundnut variety TMV2 and a net profit of Rs. 183/ha. In addition, it helped to supress the weeds and kept the soil in good tilth for most part of the year. The performance of groundnut as an intercrop in coconut, though beneficial, was not comparable to that of tapioca, colocasia or even rice as reported earlier (Table 16) (Kannan and Nambiar, 1976). Leela and Bhaskaran (1978) evaluated groundnut variety TMV2 as an intercrop in a middle-aged coconut garden and gave it the recommended packages of practices. It was reported that a yield of 1,326 kg of pods/ha was obtained, which was equivalent to a normal crop, if it was considered that land occupied by it was only 72%. Intercropping with groundnut enabled the skipping of certain tillage operations of the main coconut crop like two ploughings and liming. When the cost of these operations was also taken into account the net additional returns were Rs. 2,773 per hectare. In addition, it was observed that groundnut suppressed weeds and left the soil richer in nitrogen, phosphorus, potassium and calcium.

Mixed Cropping

Although many perennial tree crops are grown along with coconut, no yield data are available. In these cases the suitability/compatibility of the crops or their management and production values are also not known.

Mixed cropping of coconut with cacao has been evaluated (Nelliat et al., 1979). Cacao is a shade-loving tree. It was observed that cacao plants (variety Criollo), when receiving sufficient sunlight under coconut, made satisfactory growth (CPCRI, 1969) and flowered. However, cacao plants, which did not receive sufficient light, were not vigorous.

It was observed by Nelliat et al. (1979) that, when a 14-year-old coconut plantation was mixed-cropped with cacao (variety Forastero) in 1970 and the area was irrigated and cacao also received fertilisers, while there was 113% increase in the production of coconut alone, the coconut production increased by 126% and 164% with single and double row of cacao, respectively, in 1978. Apart from the increased production of coconut, 652 to 801 kg of wet cacao beans per hectare were produced on an average in a period of four years (1974 to 1978) in single and double rows of cacao-mixed crop, respectively (CPCRI, 1979). With 350 and 600 cacao plants/ha in single and double rows of cacao, respectively, the yield of wet beans per cacao plant was more with single row of cacao (3.7 wet beans/plant) than with two rows of cacao (1.3 wet beans/plant). It was thus observed that coconut and cacao were very compatible with each other. These trends have been confirmed in the same experiment in subsequent years also (Kailasam et al., 1983). For example, it was observed that the yield of cacao in 1983 was 3,920 and 4,410 kg of pods/ha in single and double rows of cacao, respectively.

Multistoried Cropping

Yields of a multistorey cropping experiment with coconut have been reported by Nelliat (1979b). Three crops—pepper, cacao or cinnamon (one and two rows) and pineapple, were interplanted with 20-year-old coconut palms (7.6 × 7.6 m) having a height of 10 m in 1972. There were 350 and 600 plants of cacao or cinnamon in single and double rows and 3,500 pineapple plants per hectare planted every three years; pepper was trained on coconut trunks and allowed to canopy up to a height of 8 m. All the crops received fertilisers individually as per recommended doses and were irrigated. Coconut, pepper, cacao or cinnamon, and pineapple were highly compatible, as the yield of the main crop, coconut, was not only maintained but very much enhanced by 75 to 122% over a period of eight years in the case of cacao multistorey model and by 90 to 120% for the same period in the case of the cinnamon model (Table 18). Yield of cacao also increased over a period of time, being much higher in 1981 and 1983 as compared to 1978. A double row of cacao decreased the yield of both pepper and pineapple. In 1983 it was concluded that cinnamon as a mixed crop was inferior and so was pineapple under cacao (CPCRI, 1984).

It was calculated in 1978, that the output from one hectare of multistoried coconut would be 17,000 coconuts, 300 kg dried cacao beans, 60 kg pepper and 4,000 kg pineapple; the cultivation expenses were Rs. 7,200 and at the prevailing market rates the net annual income could be over Rs. 20,000 (Nelliat, 1979b). It was reported, that due to the three crops (cacao, pepper and pineapple) the net added return was Rs. 9,700 and Rs. 9,100, respectively, in single and double row cacao mixtures (Das, 1982). The lower returns from the double row cacao model (even though cacao yield was more) were attributed to the much higher expenditure incurred in maintaining it.

It has been observed that the cacao crop enriches the soil fertility (Varghese et al., 1978b) by leaf fall (818 and 985 kg of oven-dry leaves in single and double row cacao models of mixed cropping). Cacao leaves are reported to contain 2.84% N, 0.26% P and 1.73% K, (Earnstman, 1968). It is further observed (Varghese et al., 1978b) that while the soil of unirrigated coconut alone contained 0.32% organic carbon in the top 30 cm soil, irrigated coconut soil contained 0.56% organic carbon, the top 30 cm soil in coconut multistorey cropping with cacao, pepper, pineapple and one and two rows of cacao contained 0.59 and 0.64% organic carbon,

Table 18: Yield of different crops under coconut multistoried cropping

Yield parameter	Cropping combinations with coconut				
	PCa1Pi*	PCa2Pi	PCi1Pi	PCi2Pi	Ci2
1. Yield of coconut (nuts/palm/year)					
– Pre-experimental** (1969-72)	45	48	32	40	41
– Experimental					
1978***	108	94	74	88	89
1981**	106	91	82	81	88
1974-81 (8 yrs. av.)**	100	84	76	76	90
% increase over pre-experimental					
1978	140	96	131	120	117
1981	136	90	156	103	115
1974-81 (8 yrs)	122	75	138	90	120
2. Yield of cacao (pods/tree/year)					
1978	14	9	–	–	–
1981	52	33	–	–	–
1983****	59	35	–	–	–
3. Yield of pepper (gm/standard/year)					
1981	700	367	1598	1087	–
1983	1237	687	1469	1604	–
4. Yield of cinnamon (gm/plant)					
1981 (quills)	–	–	162	85	115
1981 (quilling)	–	–	115	49	36
5. Yield of pineapple (kg/ha/year)					
1983	777	102	1196	565	–

** *Source*: CPCRI, 1982

*** *Source*: Nelliat, 1979b

**** *Source*: CPCRI, 1984

*P = Pepper; Ca = Cacao; 1,2 = single or double row; Pi = Pineapple;
Ci = Cinnamon

respectively. Thus it is clear that coconut multistorey cropping—perhaps more due to leaf fall of cacao, builds up soil-organic carbon.

Products

There is perhaps no other tree which has more products and uses than coconut. It is referred to as *Kalpa Vriksha* in Sanskrit meaning a tree which meets all of one's needs. In simple terms it is called the "Tree of Wealth". A Sri Lankan proverb says that the coconut has ninety-nine uses, but that the hundredth will still undoubtedly be found. Some of the most important products and uses are mentioned here (Grimwood, 1975).

The wood of the trunk, though not very durable, is used for building houses, bridges, cattle sheds, aqueducts and for timber. Formerly, the wood was also used for building ships—even up to a displacement of 120 tonnes. The bark contains resinous and waxy materials and the roots contain a drug recommended for diarrhoea. The dry leaves are used for thatching, and for making baskets, hats, mats and curtains. The petioles and midribs are used for fence posts, canes, brooms, needles, pins, fish hooks and many other things. The bud or heart of the apex of the stem, and the unopened leaf buds are eaten as salad and vegetable. The spathes are used as ladles, and also as torches when dipped in oil or resinous material.

When cut, the flower shoots or unopened inflorescence yield a liquid called toddy or *tuba*. The liquid ferments spontaneously, to give palm wine—a common alcoholic drink. Palm wine can be distilled to produce a liquor often known as *arrack*. Unfermented toddy can be used to make palm sugar or jaggery.

The main use is made of the coconut itself. Coconut oil is used among other things, in soap, margarine, biscuit and cooking oil industries. Minor quantities of oil are used in food delicacies, lubricants, cosmetics, salves, shaving creams, shampoos, dentrifices and lighting.

The hard shell, or endocarp, is used as fuel, and for making vessels, containers, ornaments, ladles, buttons, charcoal and active carbon. Destructive distillation of shells yields 35-37% charcoal, 35-37% pyroligneous acid, 6-9% settled tar and 17-20% gases. The pyroligneous acid contains a remarkably high percentage, usually over 10%, of acetic acid; about 3% of methanol is also recoverable. The tar contains 70% phenols (Mono-, di- and triphenols). Ground coconut shells are now a commercial product, used in thermoplastics. When the shell and husk are burnt in kilns, the ash is used as a fertiliser. It was once used as a source of alkali in the manufacture of soap.

The fibrous husk yields a fairly important fibre commercially known as coir, which is used for making cordage, mats and stuffing. The coir fibre is very resistant to the adverse effects of sea water. Even coir dust, the light fluffy material obtained during the separation of the fibre, is used for reclamation of land or for the improvement of sandy soils.

Mixed Farming

Mixed farming in coconut would be a silvopastoral practice. In India it would involve establishing the fodder grasses and crops; their harvesting and feeding to the milch animals and other livestock; keeping, feeding and maintaining milch animals; and recycling the cattle manure in a coconut-fodder mixed stand. It may be noted that grazing of milch cattle or livestock is not envisaged. It is reported to be possible, provided there is sufficient land for grazing. In India, due to the small size of holdings, grazing by cattle or livestock is not practicable. Grasses and fodder crops may not compete for light and air with much taller coconut palms, but they compete for soil moisture and plant nutrients. Screening trials of grasses and legumes under the shade of coconut trees gave very promising forage yields. Gautemala grass (*Tripsacum laxum*), *hybrid napier* (Pusa giant and NB21 varieties), and Guinea grass (*Panicum maximum*) gave a yield of about 50 to 60 tonnes of green grass/ha under coconut; and the legumes, Brazilian lucerne (*Stylosanthes gracilis*) and cowpea (*Vigna catjang*) gave a forage yield of about 30 tonnes/ha (Nair, et al., 1976; Sahasranaman and Pillai, 1976). At a feeding rate of 30-40 kg green fodder per animal/day, with grasses and legumes in the ratio of 3:1, an area of one hectare could support four milch animals.

The results of a mixed farming experiment conducted for six years in a well-managed 60-year-old rainfed coconut plantation on 1.04 ha of land have been reported (Mathew and Shaffee, 1979). There were 191 mature coconut trees and 42 under-planted young coconut trees. Hybrid napier grass (Pusa Giant variety) and *Stylosanthes gracilis* were planted as fodder crops. Tillage, fertilisers and manures were applied as needed by the coconut and fodder crops. Summer irrigation, using sprinklers, was provided to ensure steady growth of fodder throughout the year. Along the border and field bunds, intercrops like banana, tapioca and vegetables were grown. Black pepper was also trailed on some coconut trees. Jersey crossbred (first lactation) cows were used to start the experiment. The male calves were disposed off at the earliest opportunity; the female calves were reared till the first lactation and if found to be good milk yielders, they replaced the least economic older cows. Detailed accounts of expenditure (capital investment and operating expenses) and income (receipts and returns) were maintained for economic analysis. It was observed that mixed farming called for an extra investment of Rs. 39,790 on the capital account and Rs. 14,637 per annum in the operating account. This extra investment gave an extra income of Rs. 22,763 which included Rs. 12,000 towards the extra labour provided by the farm family. The benefit : cost ratio due to the mixed farming was 1.28 (Mathew and Shaffee, 1979). As against the annual labour requirement of 150 mandays (Nelliat and Krishnaji, 1976), the total employment for mixed farming was about 800-850 mandays, excluding the employment potential for planting/replanting of grasses and legumes which was estimated to be 150-200 mandays. It was also observed that the average yield of coconut under mixed farming actually increased from 64.7 to 69.5 nuts/palm/year during six years. In coconut alone the average yield decreased from 60.8 nuts/palm/year to 51.2 during the same period. The cowdung could also be first used as a source of biogas and then as manure.

It may be noted that not only did mixed farming give about 28% more returns but the net income was much more in quantitative terms as well, leading to greater prosperity of the farmer.

Benefit : Cost Ratios

The benefit : cost (b:c) ratios for intercropping were first reported by Nair et al., 1974. Subsequantly Nair (1979) reported the b:c ratios for rainfed and irrigated intercropping and also irrigated multistorey cropping. Under rainfed conditions a pure plantation of coconut was reported to give a b:c ratio of 2.63; all the other intercropping systems with it gave less b:c ratio though they gave much more net income, for example, with cassava 2.15; with elephant foot yam 2.18; with sweet potato 2.22; and with greater yam 2.27. In the case of irrigated coconut, the b:c ratio was 3.00; intercropping with irrigated coconut gave the same or slightly higher b:c ratio, for example, with pineapple 3.14; with cacao 3.01 and with black pepper + cacao + pineapple, 3.05. The net income in all the intercropping systems was much more due to higher investments (Nair, 1979).

Improvement of the Practice

Certain experimental innovations have been described above. There is a considerable scope to improve the practice. No information is available on the protective and service aspects. It could be presumed that with intercropping or multistorey cropping the land may have better protection from erosion. Long-term effects on soil fertility and productivity are yet to be investigated.

The most important aspects which need to be investigated are the selection of compatible crops, spacings, water management, fertiliser management and soil fertility, pest and disease management, and socio-economic aspects on a long-term basis.

4.2 INTERCROPPING WITH ARECANUT

Distribution

India is the leading arecanut (*Areca catechu*) producing country (75% of the total) in the world, producing about 249,300 tonnes of nuts from about 208,400 ha of land in 1990-91. Arecaunt grows mainly in the states of Kerala, Karnataka, Tamil Nadu, Maharashtra, Assam, Meghalaya, and West Bengal. Kerala, Karnataka and Tamil Nadu account for about 62% of area and 64% of the production in India. Another important state is Assam, which accounts for about 30% of area, and 28% production of arecanut in India (MOA, 1992; Bavappa, 1980; Vellapan and George, 1982). It is the interrelation of temperature, rainfall and humidity which determines the territorial boundaries of the arecanut area in altitude and latitude. Although it grows at an altitude of up to 1,000 m amsl, it is not at all productive in the upper limits of the altitudinal range. Further, even though it is a maritime species, it grows and thrives in areas of up to 400 km inland from the coast of India.

Edaphic Requirements

Areca palm is unable to withstand extremes of temperature and its diurnal variations (Nambiar, 1949). The range of temperature at which it can flourish is 14-36° C. It grows in areas having a very wide range of temperature—minimum 4° C (in West Bengal) and maximum 40° C (in Kerala and Karnataka). It grows in areas of high rainfall. For good growth, well-distributed annual rainfall of over 2,250 mm appears to be necessary.

It can stand rainfall of up to 4,000 mm provided the soils are well drained. It is also grown in areas of low rainfall (750 mm in Tamil Nadu) but has to be irrigated (Bhat and Khader, 1982). Bright sunshine for seven to 45 days is needed for drying the whole fruits or processed nuts.

Arecanut is grown in a variety of soils such as laterites, red loams and alluvial. The soils have to be deep and well-drained without a high water table. Being highly susceptible to sun scorch, the areca palm needs adequate protection from exposure to the southwestern sun. This protection is provided either by hillocks or by tall evergreen trees. It is for this reason that the traditional arecanut gardens are located in the valleys of hill slopes which are protected by forests growing all around. In areas where adequate natural protection is not available, the required condition is created by growing trees such as coconut, mango, *Sesbania aegyptica*, etc. (Bhat and Khader, 1982). To prevent sun scorch, the north-south line may be deflected by an angle of 35° towards the west and a shade crop of banana is raised with advantage which, in addition to giving protection to the seedlings from sun scorch, also gives good returns.

Like coconut, due to its height, the particular shape of its crown and wide spacing of planting, the arecanut palm permits the growing of annual, biennial and perennial crops under it. The pattern of the crown permits passage of sufficient sunlight for the growth of intercrops. The areca palm is planted at a minimum spacing of 2.7 × 2.7 m. It is reported, that 88.4% roots occurred up to 1 m away and around from the trunk (excluding up to 20 cm diameter); 80% roots occurred within 60 cm soil depth and 97.4% within 90 cm soil depth (Murthy and Pillai, 1982).

Crops

Banana, cacao, black pepper, pineapple, betel vine, elephant foot yam, tapioca, paddy, sorghum, cowpea, vegetables, turmeric, ginger and Guinea grass are grown successfully under arecanut in inter- and mixed crops. Although the crops grown vary from tract to tract, a crop like banana may be universally grown. The areca growers of Assam prefer citrus; the growers of Kerala prefer *Dioscorea*, yams and tapioca; the growers of Karnataka prefer cardamom, pepper, banana and yam (Khader, 1982). The crop chosen depends on several factors such as the agroclimatic conditions, demands and preferences of the local market and the needs and preferences of the farmer. Although the practice of mixed cropping a perennial tree/shrub crop with another perennial tree crop is old e.g. citrus, jack fruit and coconut with arecanut, the introduction of cacao is recent (Bhat, 1979).

Arrangement of Components

In arecanut nurseries, banana is grown to provide shade to the seedlings. In the case of mixed cropping with black pepper, areca palms of more than 10 years of age, and 7-8 m tall serve as good standards (Nayar, 1982). In north West Bengal, the farmers usually grow one crop at a time. This is mainly due to the lack of technical knowhow, limitation of irrigation, constraint of availability of planting materials, etc. In the case of banana, besides the main crop, two ratoon crops can be taken and then the crop will need to be replanted in pits dug at changed sites.

In terms of space, the areca palms are planted at a spacing of 2.7 × 2.7 m preferably with the north-south line being deflected by an angle of 35° towards the west. The outermost area of the palms on the southern and southwestern sides can be protected from the scorching sun by growing tall and quick-growing shade trees or by covering the exposed areca stems with areca leaves or leaf sheaths (CPCRI, 1986a).

Banana and cacao are grown at a spacing of 2.7 × 5.4 m which is equal to the rate of one banana or cacao plant for every two areca palms; elephant foot yam (*Amorphophallus companulatus*) is planted at the same spacing as the areca palm, namely, 2.7 × 2.7 m; in the case of pineapple, trenches 60 × 120 × 60 cm deep are dug between two areca palms and in each trench threee suckers of pineapple are planted; Guinea grass (*Panicum maximum*) is planted in two rows, 50 cm apart between two rows of areca palms and/or on the edge of drains; about 800 to 1,000 areca palms per hectare are used for growing pepper as a mixed crop (Khader, 1982). In the case of pepper, 30 cm cube pits are dug, 0.75 m away from the base of the areca palm, preferably on the northeastern side (Nayar, 1982).

Resources Input and Utilisation

Since major arecanut areas are situated in the regions of heavy rainfall, the soils are severely leached, often eroded and poor in fertility. On the other hand, a steady and high yield of perennial crop like arecanut will need adequate supplies of plant nutrients in the soil throughout its life. Therefore an annual application of 100 g N, 40 g P_2O_5 and 140 g K_2O in the form of chemical fertilisers and 12 kg green leaf and compost per bearing areca palm are recommended. For high-yielding variety "Mangala", the application of 150:60:210 g N, P_2O_5 and K_2O, respectively is recommended. Half the dose of the fertilisers is applied before the onset of the southwest monsoon, and the remaining half dose of fertilisers, and a full dose of green leaf are applied in September-October. For young plants, the full dose of green leaf and compost is applied from the first year of planting itself; one-third, two-third and the full dose of recommended fertiliser is applied during the the first, second and third year of planting, respectively. In case of drought, the arecanut palms need to be irrigated once in three to five days depending on the soil type and season (CPCRI, 1986a). All the intercrops should receive manuring according to their own needs (Khader, 1982). When black pepper is grown with arecanut as a standard, the manure and fertiliser dose for arecanut should be doubled. Application of lime at the rate of 500 g per standard in alternate years is reported to be beneficial (Nayar, 1982).

The arecanut palm is attacked by a few pests (such as mites, spindle bug, inflorescence caterpillar, root grubs), major diseases—such as fruit rot and bud rot (*Phytophthora aracae*), fruit rot (*Ganoderma lucidim*), band or

Hidimundige disease, and dieback of inflorescence (*Colletotrichum gloeosporioides*), and a few minor diseases such as stem bleeding, bacterial leaf stripe, and nut splitting. It has a few nursery diseases such as yellow leaf spot and wilt. All these need to be managed and controlled carefully and need considerable inputs. Similarly all the inter- and mixed crops have to be provided with their respective plant protection measures (CPCRI, 1986a; Devashayam and Nair, 1982; Saraswathy and Prakasam, 1982).

Arecanut requires high initial investment, like any other perennial plantation crop. The first year cost of establishment is as high as Rs. 39,370 per hectare; there is an investment of Rs. 18,000 to 19,000/ha/year during the next three years, totalling an investment of Rs. 95,490 per hectare in four years. The crop starts yielding from the fifth year and stabilised yields are achieved from the seventh year onwards. The estimated costs and returns reveal that all the costs are recovered in 12 years. From the 13th year onwards a net profit of at least Rs. 16,000 per hectare could be expected annually on 1983-84 prices (CPCRI, 1986a).

It is reported (Singh and Roy, 1983) that in a well-managed arecanut garden, for maximum returns, in north West Bengal, as many as 360 days of manual labour are needed, constituting about 43.3% of the total cost. The next most costly item is manures and fertilisers, which accounts for 32.4% of the total cost.

Information about labour input for black pepper grown as a mixed intercrop with areca palm indicates that during the first year 103 man days of labour are needed; this is 85, 155, 204 and 194 man days in the second, third, fourth and fifth years after planting, respectively. It remains at a constant figure of 194 man days per annum from the sixth year of plantation (Nair, 1982).

Yields

A large number of annuals have been evaluated as intercrops with arecanut, in Kerala, Karnataka, West Bengal and Assam by the Central Plantation Crops and Research Institute, Kasargod (CPCRI).

It is observed that the yield of arecanut did not decrease due to intercropping; on the other hand it increased when intercropped with ginger, betel vine, black pepper, elephant foot yam, and arrow root among others. The average per hectare expected yields of some intercrops reported (Khader, 1982) are: banana, 4,500 kg; elephant foot yam 12,000 kg; pineapple 8,000 kg; Guinea grass 18-22 tonnes in five to six cuttings; cacao dried beans 800-1,000 kg. Depending upon the market prices, the practice of intercropping is highly profitable (Table 19). Elephant foot yam, tapioca, *Dioscorea*, arrow root among the tubers, banana and pineapple among the fruits, ginger and pepper among the spices, and betel vine, are all paying intercrops. The yields and profitability vary due to the agroecological conditions, the varieties tried and the management practices adopted. Only ginger has proved unprofitable in north West Bengal. It is reported (Sannamarappa *et al*, 1982) that due to continuous intercropping, yields of ginger and turmeric decreased continuously over a period of four years.

Table 19: Production and profit due to intercropping under arecanut

Intercrop	Location	Yield (kg/ha)	Net profit (Rs./ha)	Source
Tubers				
Yam	Vittal	12000	1550	Bhat 1974
Elephant foot yam	Palode	6434	1700	Abraham 1974
	Kannara	4245	1398	Sadanandan 1974
Tapioca	Palode	10246	1851	CPCRI 1977
Dioscorea	Palode	6744	1824	Abraham 1974
Sweet potato	Palode	712	61	Abraham 1974
Arrow root	Palode	4000	1100	Bhat 1974
Fruits				
Banana	Vittal	4000	1650	Bhat 1974
	Kahikuchi	12200	728	Roy 1974
	Mohitnagar	–	850	Brahma 1973
	Mohitnagar	3175	2850	Singh et al. 1982
	Hirchalli	–	1000	Nagaraj 1974
Pineapple	Vittal	8000	850	Bhat 1974
	Kannara	1953	703	Sadanandan 1974
	Palode	3942	847	Abraham 1974
	Hirchalli	–	500	Nagaraj 1974
	Kahikuchi	15000	2379	Roy 1974
	Mohitnagar	3202	737	Singh et al. 1982
Spices				
Ginger	Kannara	2650	905	Sadanandan 1974
	Kahikuchi	9800	998	Roy 1974
	Mohitnagar	823	–2033	Singh et al. 1982
Pepper	Vittal	1300	8100	Nair 1982
	Mohitnagar	415	6504	Singh et al. 1982
Betel vine	Mohitnagar	648400 Nos	3294	Singh et al. 1982

Although banana is one of the more popular intercrops, there is evidence to indicate that the continuous cultivation of banana for a period of 12 years reduced the yield and weight of arecanut. However, if the banana was managed for varying periods, the yield and weight of arecanut were not affected. For example, as compared to the number of nuts per palm (145) and wet weight of nuts per palm (5.24 kg) obtained from a pure crop of arecanut, the number of nuts per palm was 103, 136 and 183 when banana was grown with arecanut for 12, six and three years, respectively. The corresponding value for the wet weight of nuts per palm was 3.83, 5.14 and 6.62 kg, respectively (Sannamarappa and Muralidharan, 1982). Although reduction in the yield of arecanut has been reported earlier also (Roy,1974; Nagaraj,1974; Bhandary,1974), the practice still remains profitable due to the very high yields of banana. "Mysore Poovam" variety of banana has given the highest yields (CPCRI, 1981).

Mixed Crops

A few perennial trees/shrubs/climbers are grown as mixed crops with arecanut. Mixed cropping of arecanut with coconut, citrus, jack fruit and other perennials in Karnataka was reported in as early as 1893 (Watt, 1893). However, the scientific endeavours to evaluate mixed cropping in arecanut are of recent origin—1964 (Bhat, 1979). It was realised by 1968, that cacao was likely to do well as a mixed crop (Bhat and Leela, 1968b). Cacao is one of the main perennial crops which has been studied for long. At present it is considered to be the most important mixed crop with arecanut. Some of the other perennial crops studied are black pepper, cardamom, clove, nutmeg, betel vine and coffee (Sannamarappa and Muralidharan, 1982).

In an experiment, in which arecanut and cacao (variety Criollo) were planted in the ratio of 1:1 in a quincunx method adopting a spacing of 4 × 4 m the data indicated that the mean yield of arecanut in the mixed crop was more in the number of nuts per hectare (at 185,000) than found in the pure crop of arecanut (at 134,000), but less in the wet weight of nuts (at 1,482 kg) than in the pure crop of arecanut (at 1,869 kg). However, since the mixed crop yielded 1,225 kg of dry cacao also it was more profitable than arecanut alone. Based on the market prices, the gross income from the arecanut and cacao mixed planting was Rs. 18,949/year, as compared to Rs. 13,083 from the pure arecanut plantation (Bhat, 1979). (For further details on cacao refer to section 3.3).

Mixed cropping of arecanut and black pepper for a period of 10 years showed that there was no adverse effect on the yield of arecanut (average number of nuts per palm was 307 and 310 with and without pepper, respectively; the values for the fresh weight of nuts per palm being 10.1 and 10.4 kg, respectively). However, pepper helped to augment the income by Rs. 8,940 per hectare per year (Nayar, 1982).

In the case of arecanut and cinnamon mixed cropping in the ratio of 1:1, the yield of arecanut increased to 24,445 kg per ha from 17,892 kg per ha within a period of three years (Sannamarappa and Muralidharan (1982).

Use and Products

Arecanut is traditionally used as a masticator in India, and in the south and southeast Asian countries. Its ethnoreligious importance, particularly in India, is unique. It is chewed in *paan* (betel leaf) or independently. The nut contains fat, polyphenols (tannins), alkaloids, polysaccharides, small quantity of protein, and vitamins B6 and C. The fat content is 14 to 15% and is extractable with organic solvents. The fat is rich in myristic acid, hence a good indigenous source of myristic acid and its derivatives. The fat can also be blended with cacao butter or coconut oil in the manufacture of confectionary and imitation dairy products. Polyphenols (tannins) form the major constituent of arecanut (38 to 47% in tender nuts and 16 to 22% in ripe nuts). Tannins can be extracted and used as a natural colouring agent in foods and also in the leather industry. The arecaunt alkaloids—arecoline, arecolidine, arecaidine, guvacine, isoguvacine and guvacolidine—constitute about 1.6% of the total nut weight. These alkaloids have antihelminthic, opthalimic, antibacterial, antidiabetic and pharmacological uses. Arecanut husk, has great scope in many industrial applications, for example, hardboard, insulation wool, cushions, paper and paperboard and activated carbon. Areca husk contains, 18.75% furfuraldehyde and yields 5.5% furfural on distillation; it also yields xylitol. Areca leaf sheaths can be used to manufacture plyboards. Many other traditional uses of leaf sheath are made, such as, caps, eating bowls, containers for toddy, and art objects (Chempakam et al., 1982).

Protective and Production Aspects

As mentioned earlier, the areca palm is suited for intercropping. The intercrops should not accentuate the problems of exhausting the soil to the detriment of the palms and pest buildup. The intercrops, apart from being able to tolerate the shade, should be able to withstand heavy dripping from the areca palm leaves during the monsoon showers. It is obvious that the intercrops should not have competitive root systems or root zones. Banana, cacao, pepper, pineapple, betel vine, elephant foot yam and Guinea grass meet these requirements.

With respect to land management and conservation, it could be presumed that intercropping will protect the land better as the drip effect of raindrops will be minimised by the intercrops.

Socio-economic Characteristics

Raising other crops in the interspace of areca palms has been a common practice since long. The long pre-bearing age, low returns during the initial bearing stage, unexpected loss due to pests and diseases and natural calamities, and in the past, the remoteness from the markets and lack of transport, have been advanced as reasons for inducing the farmers to practice intercropping in areca gardens (Abraham, 1956; Bhat, 1974; Khader and Antony, 1968; Nagaraj, 1974; Naidu, 1959). Therefore, in the past, the emphasis on intercropping was mainly due to socio-economic reasons. As a consequence, reduction in the production of the main crop of arecanut as well as that of the intercrops, resulting from inadequate management practices and incompatibility of the crops chosen were to a great extent overlooked. At present, the above reasons are partly valid and partly the socio-economic conditions have changed for the better, for example, the general level of farming, marketing and credit facilities, roads and cooperatives, is good in the arecanut-growing areas. The emphasis has therefore shifted to intensive land use and maximum productivity.

Arecanut occupies an important place in the rural economy of north West Bengal because of its large-scale domestic use. In spite of this, the efficiency of arecanut production and marketing in the region is still low.

Even though intercropping under arecanut is profitable, yet less than 15% of the area under arecanut is intercropped in north West Bengal, where the average size of an arecanut garden is only 0.45 ha (the size of an arecanut garden varies between 0.22 and 0.69 ha). The percentage of area intercropped was more (19%) in the case of small gardens (less than 0.22 ha) than that (12%) in the case of large gardens (more than 0.69 ha) in north West Bengal (Singh et al., 1982).

Improvement of the Practice

The practice has been investigated by CPCRI, Kasargod and some of the research data have been reported above. The system has great potential with respect to increasing the production and productivity by managing various varieties, inputs, spacing, etc. of the areca palm and the intercrops. There is also a lack of information about the long-term interaction/impacts of intercropping/mixed cropping. Farmers lack information, planting materials, inputs and cash resources.

4.3 INTERCROPPING IN FRUIT ORCHARDS

Fruit trees occupy land area in the form of orchards, or small clusters. Isolated trees, like mango and *jamun* (*Syzygium cuminii*), are also found scattered in various kinds of situations including in huge compounds, on roads and canal banks. The total fruit production was 18.8 and 28.2 million tonnes in 1977 and 1990, repectively (MOA, 1992). This represents an increase in production of 51% in a period of 13 years.

Distribution

The area under fruit crops has been progessively increasing in India. The total area under fruit crops was 2.0, 2.2, and 3.3 million ha in 1977, 1979, and 1990, respectively. This represents a 65% increase in area over a period of 13 years between 1977 and 1990. As per the 1983 statistics, mango was the most dominant fruit, which occupied over 40% area; citrus occupied 9.2%, banana 12.5%, apple 6.3%, guava 6.1%, pineapple 3.4%, and other fruits 22.0%, of the total area of 2.6 million ha (MOA, 1983).

Intercropping Practice

Some fruit plants are cultivated like field crops, for example, banana, pineapple, papaya and strawberries. These themselves are grown as intercrops among trees. Among fruit crops, the trees, which take a long time to bear fruit and are spaced widely, lend themselves to the practice of intercropping, for example, mango, apple, citrus, litchi, guava, etc. All these horticultural trees have specific edaphic requirements and management practices; most of the regular orchards would be irrigated (Gangolly et al., 1957; Singh et al., 1963).

Although intercropping in horticultural trees is generally practised due to economic considerations, yet surprisingly neither detailed descriptions nor research results are available. Under these circumstances only a brief outline of the practices is described here.

MANGO (*Mangifera indica*), CHIKU (*Achras zapota*), LOQUAT (*Eriobotrya japonica*), AND POMEGRANATE (*Punica granatum*)

In the case of these fruits, rainy season vegetables, winter vegetables (with irrigation), leguminous crops like Bengal gram, berseem, peas, black gram and green gram can be grown as intercrops. In fact, any crop which does not grow too tall or does not require much irrigation can be grown. Since only 60-100 trees of mango are grown per hectare, filler trees (for example, peach, plum and guava) can be grown in between mango trees. Care should be taken not to plant them too close to the mango trees. These trees should be uprooted after about 10 years before they start competing with mango trees. Papaya (*Carica papaya*), *phalsa* (*Grewia asiatica*), dwarf varieties of banana (*Musa* spp.) and pineapple (*Anana comosus*) can also be grown, depending on the agro-ecological conditions.

ORANGE (*Citrus sinensis*), OTHER CITRUS, AND GUAVA (*Psidium guajava*)

These trees generally take seven to eight years to cover the land. Till then every year the soil needs to be cultivated and green manured during the monsoon. During winter intercrops of vegetables and leguminous crops like berseem (*Trifolium alexandrinum*) and Bengal gram may be taken. If water is available citrus commences production in five to seven years after planting. It can be intercropped for two to three years.

Intercropping experiments with lemon in Meghalaya (Anonymous, 1983) have shown that it is possible to obtain a yield of 14,000 kg pineapple in 20 months after planting, 10,000 kg cowpeas green fodder per year, and 15,500 green fodder and 9,700 kg sweet potato tubers per year per hectare of interplanted area on hilly sloping lands. The lemon orchard produced 670, 770 and 2,350 kg lemon per hectare during the third, fourth and fifth years after planting, respectively.

APPLE (*Malus sylvestris*)

Apple varieties with crab seedling root stock are widely spaced (250 trees/ha) and commence production 10 to 12 years after planting. They can be intercropped for five to six years with beans or peas. Varieties on semi-dwarf root stock which are planted close (2,500 trees/ha) and commence production five years after planting, do not lend themselves to intercropping.

It is reported that intercropping a young apple orchard with grasses and legumes under rainfed conditions for one year gave yields of green grass varying from 24,000 to 31,000 kg/ha and those of legumes varying from 24,000 to 26,700 kg/ha (Table 20). The growth of apple trees as determined by stem girth and plant height was observed to be better when intercropped with grasses and legumes. This was attributed to fertilisation of the grasses and consequent increase in soil nitrogen, contribution of underground

Table 20: Yield of legumes and grasses grown as intercrops in an apple orchard, and their impact on soil properties and growth of apple trees

Legume/ Grass	Green forage yield (kg/ha)	Increase in stem girth (cm) of apple trees	Increase in height (cm) of apple trees	Increase in soil N content (%)
Red clover (*Trifolium pratens*)	26700	2.4	110	0.010
White clover (*T. repens*)	24000	1.5	90	0.009
Lucerne (*Medicago sativa*)	25500	2.6	115	0.012
Rye grass (*Lolium multiflorum*)	31000	0.8	70	0.002
Orchard grass (*Dactylis glomerata*)	27500	0.9	68	0.002
Cannary grass (*Phalaris tuberosa*)	21000	0.7	47	0.001
Control	–	0.3	40	–0.001
SE ± M	2500	0.3	16	–
CD at 5%	6200	0.8	40	–

Source: Singh et al., 1986a

biomass to better soil conditions and control of soil erosion (Singh et al., 1986a).

APRICOT (*Prunus armeniaca*), PEACH (*Prunus persica*), PLUM (*Prunus saliciana*), AND NECTARINE

About 250 trees/ha are planted. They commence production five years after plantation. Intercrops of beans and peas are taken for two to three years.

LITCHI (*Litchi chinensis*)

Cultural requirements of litchi are similar to those of mango, except that litchi is a shallow-rooted tree and therefore deep tillage or ploughing is harmful to it. Litchi orchard needs tillage three to four times a year and must be kept free of weeds. Raising of a leguminous cover or intercrops is therefore very beneficial. Summer cover crops are especially useful for maintaining humidity. Leguminous crops like peas, beans, and Bengal gram are preferred. Young orchards could also be planted with filler trees like papaya and *phalsa* (*Grewia asiatica*).

DATE PALM (*Phoenix dactylifera*)

Intercropping is possible with the availability of water. Lentil (*Lens culinaris*), Bengal gram (*Cicer arietinum*), peas (*Pisum sativum*) and senji

(*Melilotus parviflora*), in winter and moth (*Phaseolus aconitifolius*), black gram (*P. mungo*) and green gram (*P. aureus*) in summer are recommended.

ALMOND (*Prunus amygdalus*)

In Jammu and Kashmir, saffron is cultivated in young almond plantations. Saffron corms are supplied to the farmers at 50% subsidised cost.

BEE KEEPING

Apart form intercropping, there is scope of bee keeping in the orchards. This system is practised in Jammu and Kashmir, Himachal Pradesh, and Uttar Pradesh states.

Innovations

Considering the shortage of fuelwood and fodder, an innovative silvohortiagriculture practice has been experimented upon for the rainfed Siwalik hill areas of the Punjab, Haryana and Himachal Pradesh. An irrigated class I land plot 0.11 ha (36 × 32m) was planted with *Leucaena leucocephala* (one peripheral row for fuelwood and fodder (130 plants), and one mid-plot row (34 plants) for organic mulch, 34 plants of lemon (*Citrus aurantifolia* at 3.5 m spacing as second ring after *Leucaena* inside the peripheral ring and 320 plants of papaya (*Carica papaya*) as the middle rows, and turmeric (*Curcuma domestica*) and okra (*Abelmoschus esculentus*) as understorey. The trees and crops received four to five irrigations each of 10 cm per year whenever there was a moisture stress. The soil and moisture was also conserved by mulching (5 tonnes/ha) with the leaves and twigs of *Leucaena* and other plants. The economics of the practice along with those of irrigated cropping are reported in (Table 21). This silvohortiagriculture practice provided an average net income of Rs. 17,066 per ha over a period of four years as compared to an average net income of Rs. 7,852 cropping alone. The former was 117% more than the latter (Grewal et al., 1992).

Another innovation, rather unusual, in which mandarin orange trees were grown in association with *Casuarina* has been reported from the semi-arid area of Dharwad (Alloli et al., 1991). It was observed that 13-year-old mandarin orange trees which grew next to *Casuarina* performed better (3.9 m tall, 9.4 branches, 2 m crownspread, 154 flowers per five branches, 42 fruits per tree at marble stage, and 16% fruit set) than those grown alone (3.2 m tall, 8.9 branches, 1.7 m crown spread, 40 fruits per tree at marble stage, and 9% fruit set). However the fruits grown next to *Casuarina* were of poorer quality (8.4% total soluble solids, 3.8% total sugar, and 0.3% acidity) than those grown alone (9.3% total soluble solids, 6.2% total sugar and 0.2% acidity).

There is no evidence/explanation offered as to why the quality of fruits was poorer in association with *Casuarina*. It is reported (Goor and Barney, 1976) that at the time of harvest of mandarin oranges, the soil and moisture in the case of *Casuarina* + mandarin combination was 33.4%, as compared to 8.9% with *Eucalyptus*.

Table 21: Net returns (Rs./ha), and yield of crops (kg/ha) from silvohortiagriculture practice

Farming system	Plant species	Net return (Rs.)/year				Average
		1985	1986	1987	1988	
Silvohorti-agriculture	*Leucaena*					
	i) Fuel	120 (480)*	1190 (4760)	680 (2720)	4672 (11680)	1665 (4910)
	ii) Fodder green	115 (1440)	265 (3320)	264 (3300)	298 (3730)	236 (2948)
	Papaya	14769 (7385)	8245 (4122)	14806 (7403)	2770 (1350)	10147 (5074)
	Lemon	–	100 (346)	3000 (1000)	5000 (1680)	2025 (679)
	Turmeric	–	–	2250 (750)	2520 (840)	1192 (398)
	Okra	–	7200 (6700)	–	–	1800 (1675)
	Total	15004	17000	21000	15260	17066
Agriculture-double cropping	Maize + black gram/ wheat + mustard	7624	9078	4902	9805	7852

Sources: Grewal et al., 1992

*Values in parentheses are yield (kg/ha)

With the introduction of subtropical peach cultivars (low chilling), an agrohorticulture innovation has been possible. The Floridasun variety of peach (*Prunus persica*) planted at 7 × 7 m spacing, could be intercropped up to an age of six years with cowpea, which yielded on an average 587 kg of grain/ha/year for the first three years. Sesamum was also observed to be promising during the same period. The yields declined after the peach trees were more than six years old. On an average of five years, turmeric yielded 2,420 kg rhizomes/ha/year; soybean put on good vegetative growth but did not yield any grain; pineapple though yielded on an average 556 kg fruit/ha/year for five years, yet it was uneconomical. Cowpea, sesamum and turmeric did not adversely impact the yield of the peach fruits (Arora and Mohan, 1989). The nutrient level in the peach leaves was lower with intercrops than without intercrops. While the yield of the fruit was not adversely affected in the early years, the leaf analysis data would suggest careful attention to the nutrient requirements of peach along with the intercrops.

4.4 INTERCROPPING WITH POPLARS

This is essentially a *taungya* practice, except that it is commercialised and is practised by large and prosperous farmers. The system has developed during the last 20-25 years and has been initially promoted by the match industry. Intercropping with poplars (*Populus* spp.), was started in the "Terai" belt of Uttar Pradesh. The "Terai", a submontane tract along the Siwalik and outer Himalayan hills, is characterised by deep fertile soils and a high water table. The practice is now extended to Haryana and the Punjab also. In the case of share-cropping, land is given on lease to farmers who are allowed to sow crops for two to three years after planting the trees. The land is handed over to the lessees one year before planting the poplars.

Crops

In the case of the Terai region, since the rainfall is good, irrigation is available, and the climatic conditions are otherwise favourable, a wide variety of crops is grown all the year round. During the *kharif* season (July-October) soybean, maize, grain legumes (black gram and green gram), and paddy (before planting poplars) are grown. During September-January, oil seeds like mustard are grown. During *rabi* season (November-April) wheat, potato, peas, Bengal gram and lentil are grown. In Haryana and the Punjab, rainfall is much less hence poplars are grown only on irrigated lands.

Tree Clones

Clones of *Populus deltoides* were first introduced in the Terai belt of Uttar Pradesh in 1959. *Populus deltoides* IC has been more successful than other species. In a period of nine years, irrigated trees of *P. deltoides* have been reported to attain a height of 22.5-24.8 m and diameter at breast height of 28.2-33.2 cm. Some of the clones that have been successful in farmers fields are G-3, G-48, D-121, ST-67, S7C4, S7C8 (Chaturvedi, 1992).

Arrangement of Components

The plantation area is cleared of slash and stumps and given a thorough burn. The area is then ploughed and harrowed for cultivating the agricultural crops, one year in advance of planting the trees. Before planting poplars, two agricultural crops are taken. Poplars are planted when the third crop (second *rabi* season) is already in the field. After tree plantation, the crops can be taken up to five years starting with the year in which the poplars were planted.

Resources Input and Utilisation

This practice is a commercial operation, the main requirement being to keep the area clean of weeds. Since trees are grown by big farmers for use by the match industry, they are also treated with care. Trees are planted in well-worked soils in pits 30 × 30 cm and 75-80 cm deep, spaced at a distance of 4 × 5 m, or 5 × 5 m, to permit mechanical cultivation of intercrops. Each pit receives 2 kg of farmyard manure and a few grammes of aldrex powder (5%). Prior treatment of the root system, with antifungal agents is recommended. A belt of natural forest of plantation of some other species was considered desirable around every unit of poplar plantation raised by the forest department in the Terai. With the extension of this practice in Haryana and the Punjab, where there are no forests, this requirement does not appear to be valid.

Yields

The trees are ready for harvest between the tenth and twelfth years, when the increase in volume slows down, and the trees attain a dbh of about 30 cm. At this stage the wood is considered suitable for plywood, matchwood and packing cases. A rotation of 12 years is considered appropriate, when a 5 × 4 m plantation is likely to yield 100 cu m of wood or 20 cu m of pulpwood per hectare.

A case study with G-3 clone, has indicated that the internal rate of return varies from 31.5, 30.0, 28.4, 26.5, 25.8 and 25.2% after the fourth, fifth, sixth, seventh, eighth and ninth year of planting, respectively. Alhough the internal rate of return was the highest after the fourth year, this rotation is viable only when there is local demand for small size wood. However, as the demand for poplar wood increases for plywood, the ninth year rotation could be more profitable (Chaturvedi, 1992).

With respect to crop yields, 3.5 tonnes wheat and 3.5 tonnes maize per hectare are harvested during the first year of plantation; the crop yield declines to 2.5 tonnes wheat and 3.0 tonnes maize per ha during the second year and it is only 1.0 tonne wheat per hectare in the third year.

Ralhan et al. (1993) have reported a loss of 510, 500, 1,170 and 1,020 kg of wheat grain per hectare in the first, second, third and fourth years after planting of poplars at 5 × 4 m spacing, respectively. This represented a loss of 9.9, 9.4, 23.3 and 20.3% yield in the first, second, third and fourth years after planting of poplars, respectively. It was further reported that the farmers were "amply compensated" by the sale of poplars after six-and-half years (Rs. 32,500 profit per ha). However no economic analyses has been done for the whole practice (Ralhan, 1993).

The agricultural crops and trees are reported to provide severally a positive benefit : cost ratio.

Modern agricultural practices are providing necessary flexibility to raise highly remunerative cash crops such as vegetable seeds, spices and vegetables. A sample of such crops and return per hectare are reported in (Table 22).

Protective and Production Aspects

These are similar to those of *taungya* system, with the difference that intercropping with poplars is a highly commercialised practice. This practice will get degraded if there is excessive irrigation or over-fertilisation or damage to trees due to mechanical weeding. Over-fertilisation of crops coupled with irrigation, weakens the tree roots. Physical damage to the trees at the base while tractor harrowing the land, becomes an entry point for several pests and diseases.

Poplar trees have a shallow and widely-spreading root system, and invade the area by root suckers. Toky and Bisht (1992) have reported that a six-year-old plant of *P. deltoides*, growing at a spacing of 5 × 5 m and having been watered only during the first year of plantation in a semi-arid area, had a horizontal root spread of 580 cm, as compared to the

Table 22: Income from various crops grown with poplars

Poplars with	Yield (kg/ha)	Duration (months)	Gross income (Rs./ha)	Net income (Rs./ha)
Tomato seed	–	4 – 5	7500	3100
Onion	15000	5	7500	1800
Ginger	1100	8 – 9	11000	3400
Tumeric	4000	8 – 9	14000	3900
Carrot seed	200	4	3000	1150
Turnip seed	300	4	3000	1250
Mustard	800	3 – 4	2400	1400
Soybean	2500	4	4000	1700
Peas	4000	3	5000	2200
Berseem (green)	50000	5 – 6	5000	2400
Pulses (black/green gram)	1000	3	2200	1000

aerial crown diameter of 460.7 cm. This species also has a very large number of total, primary, secondary and tertiary roots, as well as root biomass; 75% of these roots are concentrated in the top 30 cm of the soil. In view of this rooting pattern, *P. deltoides* is a highly competitive tree with agricultural crops. This practice is possible due to the high commercial value of poplar wood and also due to the high inputs of fertilisers and irrigation for agricultural crops.

Socio-economic Aspects

On the land leased by the forest department, this practice is a partnership between equals—big or well-to-do farmers who are willing to pay for lease of the land and invest in modern agriculture on the one hand, and the forest department on the other hand. In the case of private farms, the market forces are at play, as discussed earlier in section 2.3 dealing with woodlots.

Improvement of the Practice

This practice has a great scope of improvement by research. As the practice is highly input-oriented and has great scope of expansion, there is need of research on: (1) inputs, irrigation, water-management; (2) crop compatibility since poplars throw up a large number of root suckers, when intercropped; (3) socio-economic analysis to provide fair returns to all the participants, and (4) generation of information on tree yields as no factual field data are available. There is also an ongoing search for new fast-growing clones of poplars, as well as new tree species other than poplars. Of late *Anthocephallus cadamba* is being experimented on and is being introduced for the same purpose.

CHAPTER 5

FOREST BASED SILVOAGRICULTURE PRACTICES

5.1 SHIFTING CULTIVATION

Shifting cultivation is an important and very extensive silvoagriculture practice in which forest dwellers grow agricultural crops in forests.

Distribution

Shifting cultivation is practised extensively in the northeastern hill region and to some extent, in the hills of Andhra Pradesh, Bihar, Madhya Pradesh, Orissa and Karnataka states of India. It is called *Jhum* in the northeastern hill states and *Podu* in Andhra Pradesh and Orissa. The northeastern hill region comprises seven states/union territories—Assam, Meghalaya, Manipur, Nagaland, Tripura, Arunachal Pradesh and Mizoram. Considering the steep landscape, high rainfall and the isolation of the hill tribes, shifting cultivation is the most appropriate response to the prevailing agroecological situation. Shifting cultivation has come down through generations and the whole gamut of tribal society is interwoven with it for food production. The entire way of life, the training of youth, the whole social system, the ceremonies, festivals and the philosophy of life, are products of this system of economy (Sharma, 1980; Majumdar, 1980). It is remarkable that though the hill tribes in India have been exposed to the forces and and to the influence of advanced civilisation (*sic*) for about two hundred years now, their main agricultural pattern has not changed. The only difference observed is that the stone implements of prehistoric shifting cultivation have been replaced by iron tools. The following description pertains specifically to the northeastern hill region. (The details may vary from place to place, but the pattern is the same elsewhere also.) Not much has been reported about shifting cultivation in Orissa and Andhra Pradesh.

Climate

The climate is tropical to sub-tropical. Hot and cold waves are rare. In the northeast hill region, the maximum temperature varies between 16.3° C in the winter and 32.7° C in the monsoon; the minimum temperature varies between 5° C in the winter and 24.7° C in the monsoon. Humidity is high in all the seasons, rarely going below 75%. Normal annual potential evapotranspiration varies from 996.2 to 1,256.8 mm. The annual rainfall varies between 1,412.5 and 10,869.4 mm. The annual water surplus is generally over 100 mm over the whole of northeast India (IMD, 1982).

Forests

Forest types are mostly tropical wet evergreen, tropical semi-evergreen and tropical moist deciduous. Shifting cultivation is practised on land which is classified as forest land, which is 49% of the total area (255,050 sq km) of the northeastern region.

Area and Length of Cycle

The exact area presently under shifting cultivation in the northeastern hill region is not known. However, estimates prepared in 1974 by the states/union territories in the northeastern hill region (Table 23) indicated that about 2.7 m ha are used for shifting cultivation, of which about 16.9% area is sown at one point in time, and the rest is left fallow for natural regeneration (NEC, 1982). It may be noted that these are rather rough estimates, and no authentic data are available. In Orissa about 2.6 m ha are under shifting cultivation, of which 53.2% is cultivated annually; Andhra Pradesh, Madhya Pradesh and Bihar have only 148,000; 122,000; and 82,800

Table 23: Shifting cultivation in northeastern India

State/union territory	Area under shifting cultivation (ha)	Area sown at one point of time (ha)	Tribal families involved	Area cultivated per tribal family (ha)
Arunachal Pradesh	248580	92000 (37.0)	81000	1.13
Assam	498300	69600 (13.0)	58000	1.20
Manipur	100000	60000 (60.0)	50000	1.20
Meghalaya	416000	76000 (18.2)	68000	1.12
Mizoram	604030	61610 (10.2)	45000	1.37
Nagaland	608000	73540 (12.1)	80000	0.92
Tripura	220790	22300 (10.1)	43000	0.51
NE Region	2695700	455050 (16.9)	425000	1.07

Source: NEC, 1982

Figures in parentheses refer to percentage values of the total area under shifting cultivation

ha under shifting cultivation of which 5.0, 1.3 and 1.6%, respectively are cultivated annually. Therefore, compared to the northeastern hill region and Orissa, the area in Andhra Pradesh, Madhya Pradesh and Bihar is rather small (Ninan, 1992).

The estimates with respect to the area under shifting cultivation in Orissa (as well for other states) are extremely variable and perhaps more of "guesstimates". The earliest estimate by Money in 1939, put the area affected by shifting cultivation at 3.3 m ha, and was practised by 1.13 million tribals (CDS, 1939). The validity of this estimate has been questioned on the ground that these values represented a very high percentage (60) of the forest area, as well as of the tribal population (45) in Orissa. The next estimate was provided by the Indian Council of Agricultural Research in 1958, which was only 160,000 ha, practised by one million tribals. A third estimate, year not mentioned, put the area under shifting cultivation at 2.22 m ha, practised by 1.6 million tribals. The Task Force on shifting cultivation appointed by the Government of India in 1981-1982, put the area affected by shifting cultivation at 3.1 m ha, practised by one million tribals in Orissa. In 1984, an analysis of landsat imagery indicated that the area affected by shifting cultivation was 184,000 ha. It is obvious from the above, that beginning with 1939 to 1984, various agencies made wild speculations without ground truthing as well as defining the term "area affected" by shifting cultivation (CDS, 1990). For example, the area affected may have been cultivated currently, or could as well have been abandoned 5, 10, or 20 years ago. Further, any empty patch in a forest is not necessarily an area under or affected by shifting cultivation. CDS (1990) estimated that the area currently under shifting cultivation was only 13,000 ha, practised by 28,200 households, and the total area ravaged by shifting cultivation during the last 20 years was 64,800 ha. This calculation assumes a five-year cycle of shifting cultivation.

The number of years before a tribal family/community returns to the same land would depend on the availability of land within a particular community, and the population of the village. During the earlier decades this period was as long as 30-40 years, mainly due to the better fertility of the soil, and small population. Currently, the cycle has shrunk to five to seven years in the northeastern hill region, hardly sufficient for rebuilding soil fertility (NEC, 1982). With the present trend of a steady rise in the population of the northeastern hill region (the decennial growth rate during 1971-1981 being 35.04%), the cycle is bound to shrink further, threatening both the economic and ecological aspects. It would appear that out of 4,343,000 tribals (in 1971) as many as 2,125,000 tribals were actually engaged in shifting cultivation in the northeastern hill region.

Shifting Cultivation Practice

The practice of shifting cultivation described here pertains to the northeastern hill region, since this practice has been studied to some extent. The shifting cultivation practice consists of cutting down the forest/vegetation at various stages of development on hill slopes (20-40%), allowing the cut vegetation to dry, then burning it, cropping the land for one or two years and then abandoning the area in favour of a new one. The method of allotment of a site to the individual families of a village varies from tribe to tribe. In most cases it is decided by the village council (Mizo Hills), or by the village elders by a lottery system (Mainpur and Mikir Hills) (NEC, 1982). The size of a plot cultivated by a family of two adults and three to four chidren varies from 1.0 to 2.5 ha (Toky and Ramakrishnan, 1982). While the village site remains permanent, a temporary bamboo hut is built at each site for living and protecting the crops.

There are significant differences in the practices of different geographical areas (Ramakrishnan and Toky, 1978). For example, the Garos living in the Garo hills and on the lower elevations of the Khasi hills in Meghalaya, fell the entire forest. The larger boles and branches are removed from the site for use as fuelwood. In more densely-populated areas, as around Tura, the cycle is shorter, whereas in the interior of the Garo hills region, the cycle may be longer. At higher elevations of the Khasi hills in Meghalaya, destruction of the forest is partial. Instead of complete felling of the forest, only the branches of trees are lopped and placed over the land that has been prepared in ridges and furrows along slopes. The dried branches are burnt on top of the ridges. Owing to the fact that the destruction of vegetation is partial, that the land is prepared in ridges and furrows, and manure is applied, the fertility of the soil is maintained for a longer period. Under this practice, cropping is done for five to six, or even more years (Ramakrishnan and Toky, 1978). In Nagaland, the practice is similar to the one in the Garo hills except that the cultivators are aware of the need for soil conservation. Methods, such as laying tree trunks across the slope of the hill, allowing stumps of trees to remain in the cleared field, and in a few cases, the making of bunds and channels, are used. In the Mizo hills, the land is subjected to intensive clearing and burning and the cycle is short, hence the entire forest is in a highly degraded form, usually maintained only in the early successional stages (Ramakrishnan and Toky, 1978).

Crops

The seed mixture used and the length of the cycle vary considerably from region to region. Toky and Ramkrishnan (1981) have reported that in the long *jhum* cycle (30 years), at the lower elevations in Meghalaya, cereals constitute the major component of crops, usually sown with an admixture of beans and vegtables (Table 24). In the five-year cycle, paddy, leafy vegetable (*Hibiscus sabdariffa*), banana (*Musa sapientum*), tuber crops (*Manihot esculenta* and *Colocasia antiquorum*) predominate. The ten-year cycle is intermediate, with emphasis on cereals and carries much variety in leafy vegetables, fruits and tubers. The number of crops sown is reduced with the shortening of the cycle from 13 under a 30-year cycle to eight under a five-year cycle (Table 24). A multistoried crop canopy develops,

Table 24: Phytosociology of crops in the 30-, 10- and five-year shifting cultivation cycles

Crop species	Frequency (%)			Density (plants/m²)		
	30y	10y	5y	30y	10y	5y
Grains and seeds						
Oryza sativa	100	100	100	8.6	12.2	1.5
Sesamum indicum	80	35	5	1.1	0.5	0.1
Zea mays	80	25	15	1.8	1.5	0.2
Setaria italica	90	30	15	2.1	0.5	0.2
Phaseolus mungo	5	–	–	0.1	–	–
Ricinus communis	20	–	–	0.2	–	–
Leaf and fruit vegetables						
Hibiscus sabdariffa	10	20	20	0.1	0.2	0.2
Hibiscus esculentus	–	10	–	–	0.1	–
Capsicum frutescens	–	5	–	–	0.1	–
Legenaria leucantha	10	5	–	0.2	0.3	–
Cucurbita maxima	5	–	–	0.1	–	–
Cucumis sativus	10	–	–	0.2	–	–
Momordica charantia	–	10	–	–	0.2	–
Musa sapientum	–	10	30	–	0.1	0.3
Tubers and rhizomes						
Manihot esculenta	5	10	10	0.1	0.1	0.1
Colocasia antiquorum	5	5	15	0.2	0.2	0.2
Zingiber officinale	5	–	–	0.2	–	–

Source: Toky and Ramakrishnan, 1981

with perennial crops such as cassava, banana and castor, occupying the top layer, cereals the middle layer, and cucurbits and legumes the lowest stratum (Toky and Ramakrishnan, 1981, 1982; Mishra and Ramakrishnan, 1982). In the Khasi hills, at higher elevations in Meghalaya, a mixture of six crops—potato (*Solanum tuberosum)* as the dominant crop, and *Colocasia antiquorum*, sweet potato (*Ipomoea batatas*), maize (*Zea mays*), beans (*Phaseolus vulgaris*) and vegetables (*Cucurbita maxima*) are grown (Mishra and Ramakrishnan 1981, 1982).

Animals

The northeastern hill states/union territories of Nagaland, Manipur, Mizoram, Meghalaya and Tripura support 1.410, 0.182, 0.125, 0.400, 0.537 and 0.038 million cattle, buffalo, sheep, goat, horses/ponies, and pigs and other livestock, respectively. They also have 3.4 million poultry (PC, 1982). Shifting cultivators use pigs, poultry, cows and *mithuns* for food. They do not drink milk (Saha, 1980). Hunting, fishing and food gathering from nature's stock in the forests are important subsidiary sources of food. They do not use livestock for cultivation.

Resource Input and Utilisation

Apart from the required agroecological conditions, shifting cultivation is practised where there is availability of forest land, and labour of the tribals, and there is a lack of captial. Wherever there is sufficient land to support fallow but no capital, no other farming system would produce higher returns to the labour, than shifting cultivation. The system is therefore very labour-intensive requiring very little capital and other inputs. Major inputs are labour and seeds; no fertilisers are used. Burning of forest causes some changes in soil properties and fertility. For example Prasad et al. (1981) reported that after burning, the soil pH increased from 5.1 to 5.5, available K_2O increased from 219 to 570 kg/ha, exchangeable Ca increased from 7.14 to 9.46 m.e%, the organic carbon decreased from 1.32 to 1.05%, and there was no change in available P_2O_5 from 3.30 to 3.31 kg/ha. The extent of changes depends on the type and quantity of material burnt. Increase in pH is related to the increase in available K_2O and this may be temporary. After removal of the nutrients by plants and soil erosion the pH would go down. Mishra and Ramakrishnan (1983a,b) have studied similar changes in the 15-, 10- and five-year *jhum* cycle, as well as in terrace cultivation. They reported that before burning, the 30-year cycle had the maximum carbon, nitrogen and phosphorus in the surface soil (0-7 cm) and the whole profile (40 cm), and the five-year cycle the least; the 10-year cycle was either intermediate or equal to the 30-year cycle. After burning there was a distinct increase in the potassium, calcium and magnesium cations in the soil. Manure of vegetable and pig origin is used only in the upper Khasi Hills of Meghalaya (Mishra and Ramakrishnan, 1981, 1982). Power is provided by manual labour.

Tools and implements used for various operations are very simple and primitive, being made within the village, or more rarely being imported, but in either case the cost is negligible. Neolithic celts of various sizes and shapes have been found all over the Garo habitat in the hills of Meghalaya and this stone implement is very much like the iron hoe used by the Garos of the present day (Majumdar, 1980). Not much has changed with the passage of time, except that the stone tools have been replaced by bone, wood and iron tools. Some of the Kuki tribes of northeastern India are reported to have been using stone hoes till the middle of the last century (Goodwin Austen, 1875). However, it is to the credit of the tribal shifting cultivators that they have a wide variety of tools for various operations, such as land clearing, planting, weeding, interculture, plant protection, harvesting, thrashing, etc. (Singh et al., 1980).

The labour is provided by the family—two adults and two to four children. The chores to be perfomed by male and female members are well defined. The males contribute more labour for forest cutting and crop protection, the females contribute more labour for burning, cleaning, sowing, weeding and harvesting. However, all members participate in all the operations. On the whole, the males contribute 45.3% and females 54.7% of the labour needed on a shifting-cultivation farm. The peak demand for labour (35.5%) occurs during seed-bed preparation (forest cutting, burning and cleaning). Crop protection (6.9%), harvesting (7.4%), transportation (4.5%) and thrashing (4.6%) are extended over a long period (Sahu et al., 1980). Since shifting cultivation is done in hilly regions by manual labour, it is reported that the steeper the hill slope, the more the labour needed (Figure 10); this is true for the whole system as well as for individual operations (Sahu et al., 1980).

Yields

The yield of crops (kg/ha/year, oven-dry weight) under different systems of cultivation is given in (Table 25). The yield declined markedly with the shortening of the shifting cultivation cycle, this being particularly evident in the case of potato, which was the main crop in the practice under study. The yield of maize was comparable in the 10- and five-year cycles. The yield of leafy vegetables was high in the five-year cycle. Within the five-year cycle, the yield of the second-year crops was markedly less than that of the first-year crops. The yield of all crops under terrace cultivation was higher than that in the 10- and five-year cycles; only the yield of potato was less under terrace cultivation than under the 15-year cycle (Mishra and Ramakrishnan, 1981).

Tokey and Ramakrishnan (1982) observed that at the lower elevations of the Khasi hills in Meghalaya, where traditional shifting cultivation is practised (i.e., the entire forest is slashed and burned), the 30-year cycle required the maximum energy input (1,665 MJ/ha), followed by the 10-year (1,181 MJ/ha) and five-year cycle (510 MJ/ha), (Table 26). Almost all the energy input is in the form of labour; a very minor component of energy is contributed by seeds (24, 15 and 16 MJ/ha in the 30-, 10- and five-year cycles, respectively). The 10-year cycle gave the maximum energy output : input ratio (47.9), followed by the five-year cycle (46.7); the least ratio was with the 30-year cycle (34.1). This is attributable to the fact that the longer the cycle, the greater is the labour spent in cutting and felling of trees, clearing understorey vegetation, and collecting and burning; the 30-, 10- and five-year cycles needing 908, 685 and 83 MJ/ha, respectively for these operations. In terms of energy output per year at the higher elevations, the 10-year cycle is the best and the 30-year cycle the worst.

In the case of higher elevations in the Khasi hills (Mishra and Ramakrishnan 1981) where only branches of trees are cut and burnt, the shifting cultivators do not use any manure or fertiliser for the 15-year cycle, but use manure for the 10-year cycle and manure and fertiliser for the five-year cycle. In view of this, the maximum energy input is in the five-year cycle and the least in the 15-year cycle (Table 26). The maximum energy output : input ratio was given by the 15-year cycle (25.6), followed by the 10-year (9.85) and five-year (2.91) cycles. Although the maximum output of energy/unit of time was in the 15-year cycle (557,866 kcal), yet the 10-year cycle was not far behind (514,793 kcal); the five-year cycle gave the minimum energy output per unit of time. This may be because

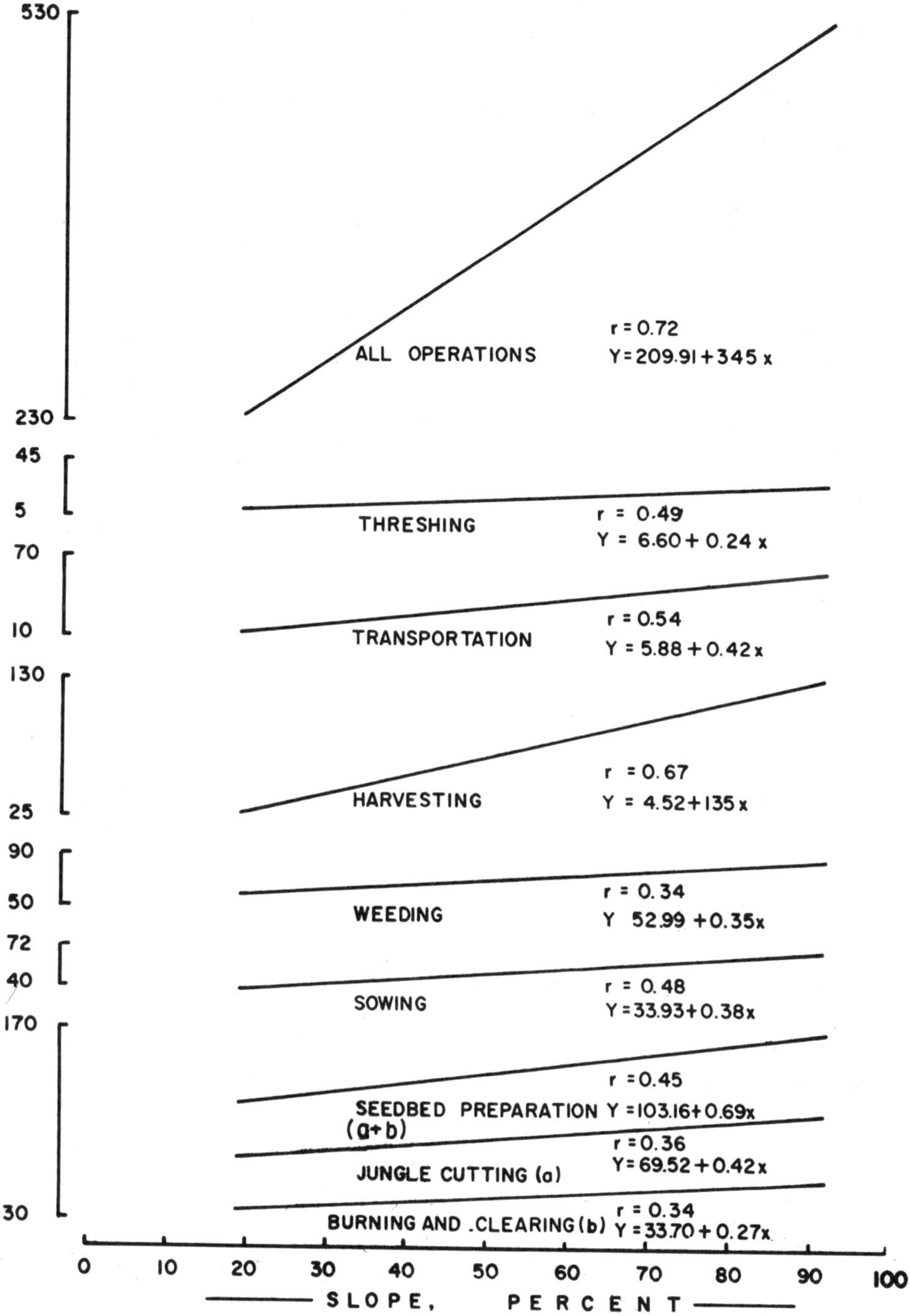

Fig. 10: Energy required for various operations of shifting cultivation at varying land slopes (Source: Sahu et al., 1980)

Note: Seed-bed preparation comprises two operations: (a) Jungle cutting, and (b) Burning and clearing

Table 25: Economic yield of crops (kg/ha/yr, oven dry wt) under different agro-ecosystems

Crop species	Shifting cultivation					Terrace cultivation (mean±S.E)	Valley cultivation (mean±S.E)
				5-yr cycle			
	Moisture content (%)	15-yr cycle (mean±S.E)	10-yr cycle (mean±S.E)	1st-yr crop (mean±S.E)	2nd-yr crop (mean±S.E)		
Root and tuber crops							
Solanum tuberosum	72.2	1580.0 (23.8)	1099.1 (5.6)	748.8 (2.3)	449.3 (1.9)	1236.5 (2.4)	–
Ipomoea batatas	70.0	446.8 (7.9)	148.3 (2.1)	–	–	–	–
Colocasia antiquorum	71.8	4.9 (0.3)	5.6 (1.9)	–	–	–	–
Cereals							
Zea mays	14.3	42.4 (1.1)	59.3 (1.9)	61.6 (1.3)	8.6 (0.3)	122.8 (1.2)	–
Oryza sativa	13.4	–	–	–	–	–	855.8 (2.7)
Legumes							
Phaseolus vulgaris	10.2	6.7 (0.3)	4.0 (0.3)	–	–	–	–
Fruit vegetables							
Cucurbita maxima	93.0	272.4 (3.7)	257.6 (2.0)	–	–	–	–
Cucumis sativus	96.5	9.1 (0.6)	6.3 (0.3)	–	–	–	–
Leafy vegetables							
Brassica oleracea	90.0	76.5 (2.1)	27.4 (1.3)	163.9 (1.4)	50.3 (1.9)	170.9 (3.1)	–
Colocasia antiquorum	80.2	2.4 (0.1)	2.8 (0.2)	–	–	–	–
Cucurbita maxima	82.0	3.1 (0.2)	2.8 (0.3)	–	–	–	–
Ipomoea batatas	80.0	10.0 (0.1)	1.1 (0.2)	–	–	–	–
Other vegetables							
Colocasia antiquorum (stem)	92.8	0.4 (0.02)	0.4 (0.02)	–	–	–	–
Cucurbita maxima (flower)	91.0	0.4 (0.02)	0.5 (0.03)	–	–	–	–
Byproducts							
Oryza sativa (straw)	10.7	–	–	–	–	–	2118.0 (4.1)

Source: Mishra and Ramakrishnan, 1981

Table 26: Energy input-output relationships in shifting cultivation

	Type of farming system					
	Shifting cultivation - cycle (years)				Terrace cultivation	Valley cultivation
	30	15	10	5		
Higher elevation system in Khasi hills*						
Energy (Kcal/ha/year)						
Input	–	340139	581200[a]	861489[a,b]	3080952[a,b]	208877[a]
Output	–	8708157	5729131	2507075	5236548	3389224
Net output	–	8368018	5147931	1646586	2155596	3180347
Output : Input	–	25.60	9.85	2.91	1.70	16.22
Net energy output per year	–	557866	514793	329117	2155596	3180347
Lower elevation system in Khasi hills**						
Energy (MJ/ha/year)						
Input	1665	–	1181	510	6509[b,c]	2843[d]
Output	56766	–	56601	23858	43602	50596
Net output	55101	–	55420	23348	37093	47753
Output : Input	34.1	–	47.9	46.7	6.7	17.8
Net energy output per year	1837	–	5542	4670	37093	47753

Source: Mishra and Ramakrishnan, 1981*
Source: Toky and Ramakrishnan, 1982**
a) 10-year cycle and valley cultivation received manure only; the five-year cycle and terrace cultivation received manure and fertiliser
b) Average of two-year cropping
c) Terrace cultivation received fertiliser; d) two crops in a year.
1 Kcal = 4.184 joule

of the heavy energy input by way of manure and fertiliser in the five-year cycle. It is not clear how the five-year cycle would respond without manure and fertiliser. On balance, a 10-year cycle at both the lower and higher elevations appears to be the best on the basis of energy output : input relationships.

Terrace cultivation on hill slopes at both the elevations had a much lower energy output : input ratio than any of the shifting cultivation cycles. However, the net energy output per unit of time was much higher than in any of the shifting cultivation cycles. Terrace cultivation receives energy for terrace construction and also in manure and fertiliser, but does not need energy for slash and burn. Valley cultivation, which receives the water, sediment and nutrients carried by the runoff, had a higher energy output : input ratio in the case of higher elevations; it had a higher ratio than the 10-year and five-year cycles, and terrace cultivation at lower elevations. In both the cases it produced the maximum energy output per unit of time when compared to any other farming system.

In terms of benefit : cost analysis, the 30-year cycle at lower elevations in the Khasi hills had the maximum value (2.13) followed by 10-year (1.83) and five-year (1.88) cycles, which were not too different from each other. However, if one takes into account the income per unit of time, the 30-year cycle was the least efficient (Rs. 99); and the 10-year (Rs. 762) and five-year (Rs. 794) cycles were efficient and comparable. In the case of higher elevation shifting cultivation in the Khasi hills, the 15-year cycle gave the maximum benefit : cost ratio (5.4), followed by the 10-year (3.7) and five-year (1.9) cycles. However, if one takes into account the income per unit of time, the five-year cycle had the maximum return (Rs. 2,619); the 15-year (Rs. 1,075) and 10-year (Rs. 1,032) cycles were similar and with less than half the income as compared to the five-year cycle.

It may also be noted that the income in the case of higher elevation shifting cultivation was of a much higher order since potato, a cash crop, was the dominant crop; in the case of lower elevation shifting cultivation, the dominant crops were cereal food grains.

Terrace cultivation, although having the same or less benefit : cost ratio as compared to the five-year cycle, gave much higher returns per unit of time. Valley cultivation was the least beneficial in terms of benefit : cost ratio and return per unit of time, mainly because it was cereal based.

Protective and Production Aspects

If shifting cultivation is practised properly, say as a 10- or 15-year cycle, it would have both protective and productive functions. For example, it is observed that early colonising herbaceus species (like *Imperata cylindrica, Eupatorium odoratum, Grewia elastica* and *Thysanolaena maxima*) seem to dominate up to the first five years of fallow; these are then progressively replaced by bamboo (*Dendrocalamus hamiltonii*) (Ramakrishnan et al., 1978) (Table 27). This species reaches its peak in terms of both frequency and density in a 20-year fallow (Ramakrishnan and Toky, 1978). In a 20-year fallow, broad-leaved species which are shade tolerant, such as *Vitex peduncularis, Schima wallichii, Terminalia bellirica, Bauhinia alba* and *Dellinia indica*, are also frequent. In the 50-year fallow, bamboo yields place to broad-leaved tree species which are more shade-tolerant (Ramakrishna and Toky, 1978). In such a case, a mixed forest of *Schima wallichii, Shorea robusta, Castanopsis indica, Garcinia cowa, Vitex penduncularis, Eugenia* spp. etc. establishes itself, but none of the species is able to achieve dominance (Table 27) (Ramakrishnan et al., 1978).

Early successional tree species have an exploitative growth strategy, while the late successionals are conserving in their growth patterns (Ramakrishnan et al., 1982). The early successional tree species of the community, (e.g. *Duabanga sonneratioides*) in general were found to allocate

Table 27: Importance Value Indices (IVI) of plant species at different periods of fallow in shifting cultivation

Name of species	Age of fallow (years)				
	0	5	10	20	50
Grewia elastica	71.2	–	–	–	–
Thysanolaena maxima	35.9	–	–	–	–
Imperata cylindrica	23.9	48.8	–	–	–
Eupatorium odoratum	30.8	53.3	–	–	–
Arundinella bengalensis	19.7	28.5	–	–	–
Ficus hispida	11.5	–	26.1	16.7	–
Panicum khasianum	10.3	12.1	–	–	–
Panicum maxima	9.6	11.2	11.0	–	–
Ensete superba	9.3	–	–	–	–
Care cruciate	–	11.3	–	–	–
Setaria glauca	–	10.3	–	–	–
Cymbopogon khasiana	–	9.2	–	–	–
Dendrocalamus hammiltonii	–	1.9	33.8	53.8	10.0
Cyperus globosus	–	–	28.8	–	–
Litsea spp.	–	–	11.4	–	–
Mesia spp.	–	–	6.9	–	–
Combretum spp.	–	–	6.6	–	–
Macaranga spp.	–	–	6.4	–	–
Mella arborse	–	–	6.5	–	–
Machillus spp.	–	–	5.4	–	–
Vitex peduncularis	–	–	–	38.5	23.0
Schima wallichii	–	–	–	35.3	32.0
Terminalia bellirica	–	–	–	33.8	–
Bauhinia alba	–	–	–	27.8	–
Dillenta indica	–	–	–	27.8	–
Keryla arborea	–	–	–	26.7	–
Melia arborea	–	–	–	–	30.0
Castanopsis indica	–	–	–	–	27.0
Garcinia cowa	–	–	–	–	25.0
Eugenia spp.	–	–	–	–	13.0
Prunus spp.	–	–	–	–	10.0
Psychjotris spp.	–	–	–	–	9.0

Source: Ramakrishnan et al., 1978
Note : Stump sprouts not included

less to the root system (Figure 11) than the late successional species, (e.g *Artocarpus chaplasha*). A similar pattern of allocation of biomass, nitrogen, phosphorus and potassium, as reported above, was noted in the early successional non-sprouting herbaceous species vis-a-vis late successional

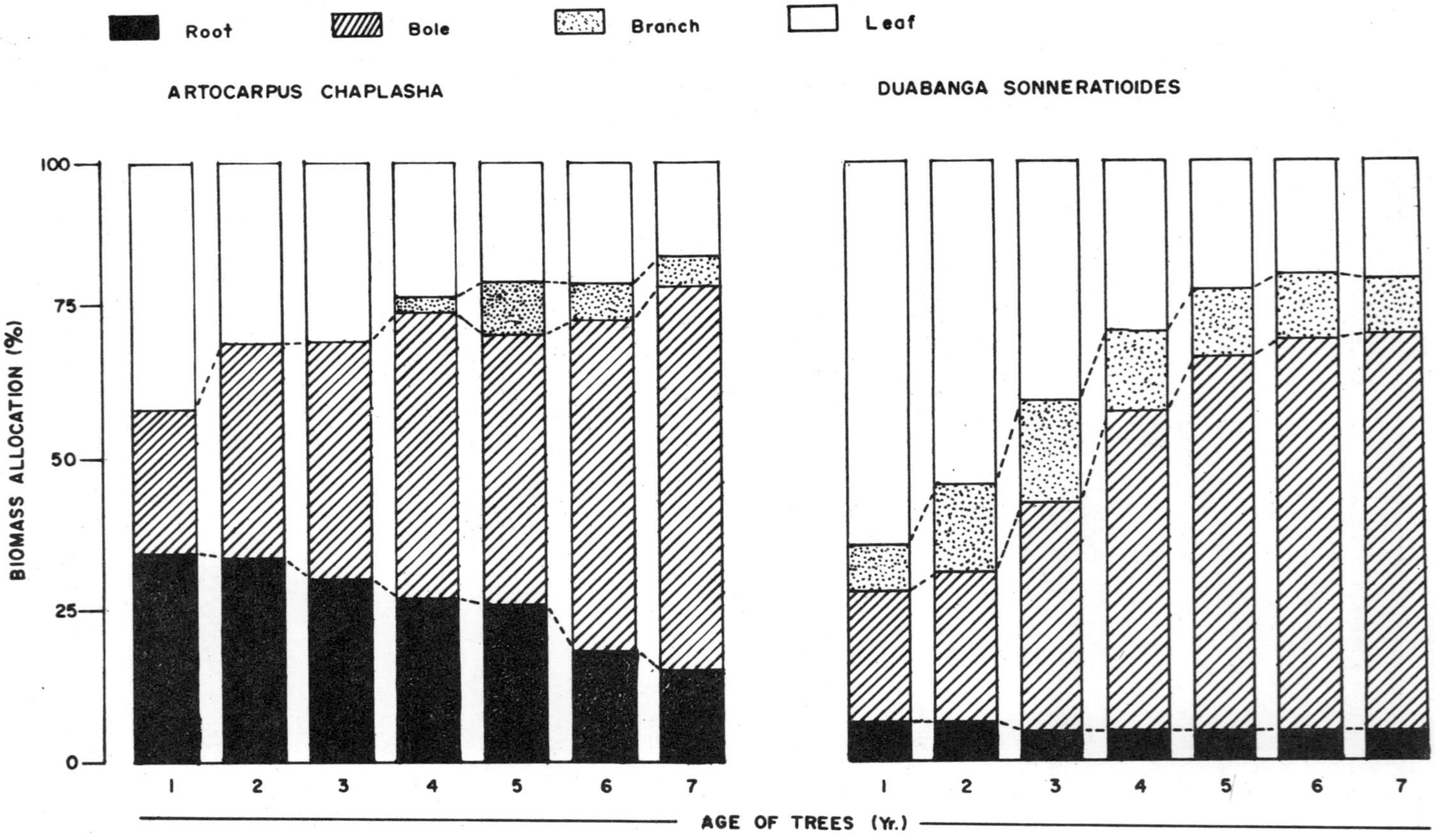

Fig. 11: Biomass allocation by early successional (*Duabanga sonneratioides*) and late successional (*Artocarpus chaplasha*) vegetation in shifting cultivation fallow (Source: Ramakrishnan et al., 1982)

herbaceous species, as well as early successional sprouting herbaceous species (Saxena and Ramakrishnan, 1982).

The shallow spreading root systems of the early successionals seek their nutrients from the richest part of the top soil (20 cm), although they provide poor support. This type of root system exploits the short-term increase in plant nutrients (and water) available in disturbed sites. On the other hand, the late successionals have higher allocation to their roots (Figure 11) in order to draw upon the nutrient pool from the deeper soil layers, and therefore exhibit deeper roots with little lateral spread. This will partly explain why the shifting cultivation cycle should be of a longer duration.

Shifting cultivators of the Jaintia hills in Meghalaya are reported to plant small branches of pine containing one or two cones at a spacing of 2 × 2 m during the course of shifting cultivation. After the area is cleared of agricultural crops, pine is allowed to grow there. In this way, about 40,000 ha of land have been rehabilitated by them (Suchiang, 1984).

Mixed cropping, accompanied by successive maturing and harvesting of crops over a long period of time, is a good protective benefit of shifting cultivation. A further advantage of the system is that the rapidly depleting/declining resource of soil fertility is utilised more efficiently and effectively by the crops in time and space. The system also provides an inbuilt "insurance" to the farmer; one of the crops is likely to give good returns even under adverse climatic conditions, when some crops may partially or totally fail.

In the early stages of the cropping cycle, shifting cultivation is a major cause of soil erosion and hence, not only it is detrimental to the land-water-vegetation resources of the region, but also, with the passage of time, it causes the system to be inefficient and less profitable. The intensity and amount of rainfall are high; the rainfall erosion index (EI 30) values reported are 228.7 (Imphal, Manipur), 407.5 (Shillong, Meghalaya), 635.4 (Gauhati, Assam), 1181.0 (Lakhimpur, Assam), 1280.6 (Agartala, Tripura) (Rambabu et al., 1977) and 677 at Barapani, Meghalaya (Anonymous, 1984). Most of the EI 30 values occur during the southwest monsoon (Figure 12). The annual soil loss is reported to vary widely from 3.3 to 201 tonnes/ha. A soil-erosion calendar relating soil loss with agricultural operations, crop canopy, rainfall (for 10 to 85% slopes) has been reported (Singh and Singh 1980, 1981). The maximum soil loss occurs during burning and clearing, and sowing and weeding of crops when the rainfall and erosion index values are high, and also when the canopy is relatively less and the movement of humans on the sloping land is at its maximum (Figure 13).

Soil erosion under the first and second years of cultivation, from an abandoned field, and from a bamboo forest is reported to be 146.6, 170.2, 30.2 and 8.2 tonnes/ha/annum, respectively. It is estimated that nearly 18.1 million tonnes of soil are eroded, and 603,130 tonnes of organic carbon, 97 tonnes of available P_2O_5 and 5,690 tonnes of available K_2O are lost annually from the land subjected to shifting cultivation (Prasad et al., 1981).

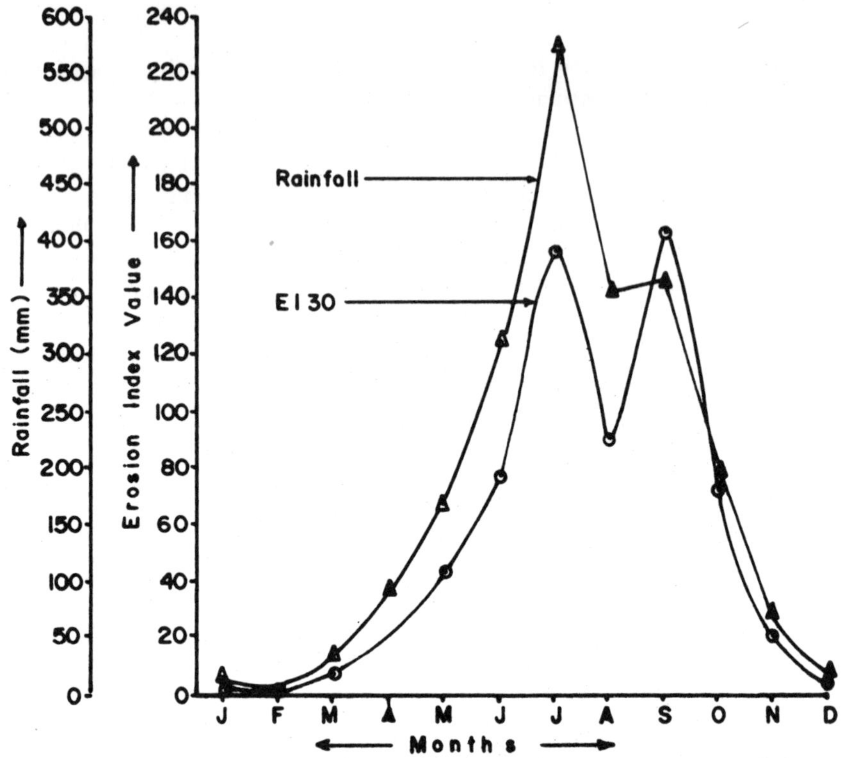

Fig. 12: Monthly rainfall and erosion index (EI 30) values at Barapani, Meghalaya (Source: Anonymous, 1984)

Mishra and Ramakrishnan (1983a) reported runoff and soil losses, rainfall percolation values, and nutrient losses under 10- and five-year shifting cultivation cycles, from one-, five-, and 10-year fallows after shifting cultivation and terraced agriculture. A comparision of the five- and 10-year cycles suggests that runoff, loss of soil, nitorogen and phosphorus were greater under the former, than the latter, although the loss of cations showed a reverse trend. This is understandable, since soil under a five-year cycle may be considered to be relatively poorer both in physical properties and in cations. All the losses are markedly reduced during the fallow period, as may be expected, due to the development of a protective vegetative cover. According to Mishra and Ramakrishnan (1983a) and Ramakrishnan (1992), while shifting cultivation cannot be sustained with a shorter cycle, terracing does not seem to offer an alternative because of the high runoff, soil and nutrient losses, the need of heavy inputs of fertilisers, and the high incidence of weeds. The adverse conclusion by Mishra and Ramakrishnan (1983a) and Ramakrishnan (1992) regarding terraces is not supported by others.

Improvement of the Practice

ICAR Research Complex for North Eastern Hill Region, along with its

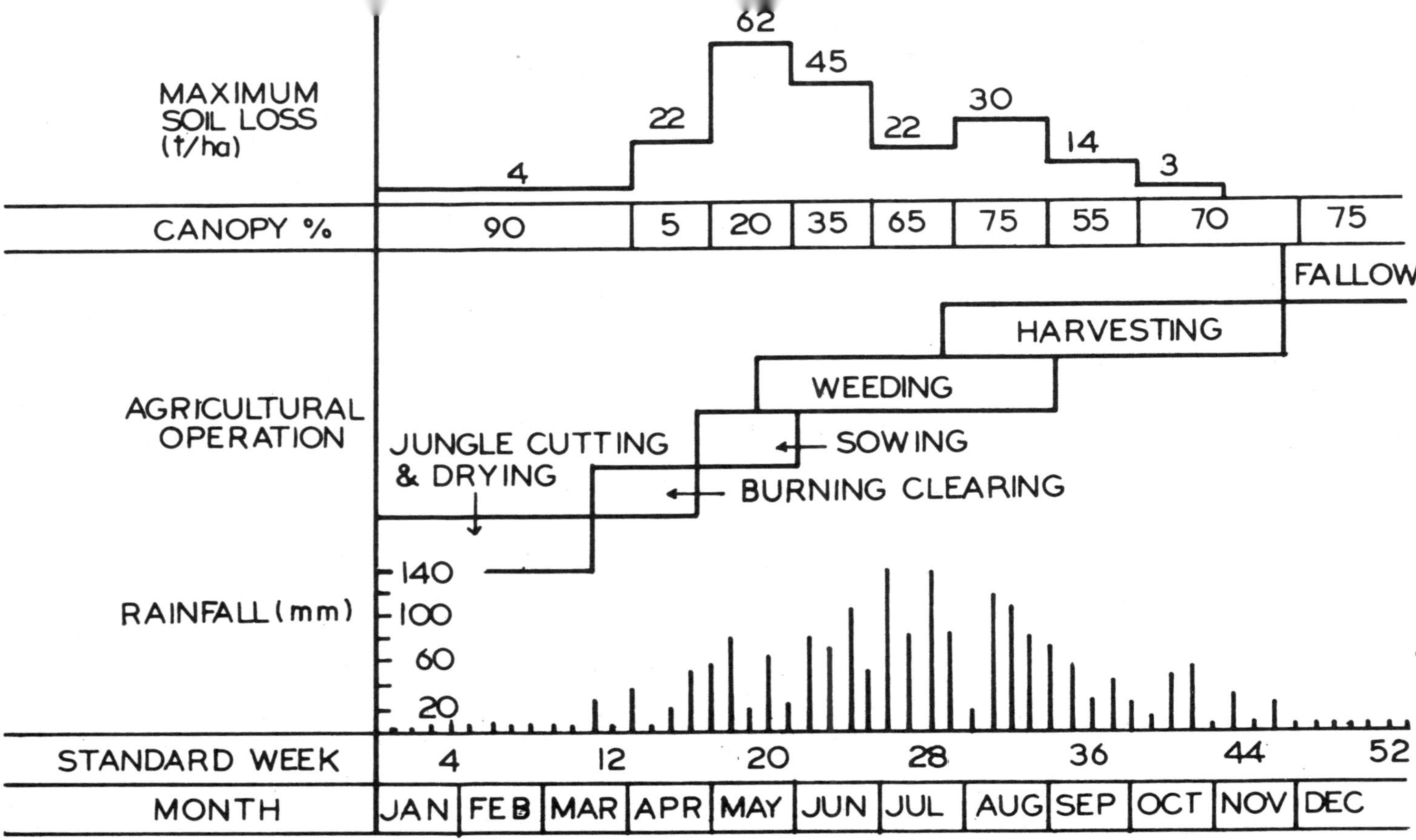

Fig. 13: Monthly soil loss during various operations of shifting cultivation (Source: Singh and Singh, 1981)

centres, is engaged in research and improvement of the system. The major biophysical weakness of shifting cultivation is the destruction of the vegetative cover and cultivation of steep slopes not suited for agriculture, without soil and water conservation measures. The ideal solution would be settled agriculture with food crops grown on gentle slopes and valley lands; fruit and forest trees grown on steeper slopes; or home gardens.

ICAR Research Complex has suggested the following silvoagrohorticulture model land-use plan as an alternative to shifting cultivation (Singh, 1981):

Part of slope	Approx. % of total area	Suggested land use	Conservation measures
Top part	35	Forestry	—
Middle part	30	Horticulture	Half moon terracing
Lower part	30	Agriculture	Bench terracing
Valley	5	Agriculture, pisciculture & ponds	Level terraces; ponds

With a horticultural crop provided in the middle of the sloping hillside, the farmer may retain a long-term interest in the land and may be prepared to practise settled agriculture. The runoff, fertile soil and nutrients originating from the upper two-thirds of the hill slopes can be used for growing paddy on the lower portion. This system would also conserve soil and water (Table 28). In the horticultural area it is suggested to plant lemon, and intercrop it with pineapple, cowpea (for green fodder) and sweet potato for fodder and tubers. Pineapple can be ratooned, and cowpea and sweet potato can be planted every year. Intercrops can be taken up to three to four years after planting lemon, till the canopy closes. In the third, fourth and fifth year after planting, 670, 770 and 2,350 kg, respectively, of lemon per hectare were obtained (Anonymous, 1983). This suggested plan still needs wider demonstration and field trials to test its adaptability, economic viability and social acceptability. A word of caution—the suggested land-use system would need a regular supply of inputs, for example, seeds, planting materials, fertilisers, plant protection materials and trained extension personnel.

In the field of horticulture, the rejuvenation of existing but declining citrus orchards, which were being abandoned due to diseases, has been demonstrated. The activity has the potential to increase the production of such orchards, in a short time and create incentives for farmers to take to settled agriculture. High density of planting pineapples on slopes (34%), with legumes as a companion crop to act as a soil conservation measure, has provided an alternative to shifting cultivation on steep slopes. This will require the development of marketing strategies.

Another alternative farming system which holds promise is the system of home gardens. This system is practised on the plain lands in Kerala

Table 28: Runoff, soil loss and nutrient loss from different farming systems in the northeastern hill regions

Particulars	Farming system		Agriculture
	Shifting cultivation	Agrohorticulture	
Conservation measure	–	Lower 1/3 area bench terraced for agriculture, and upper 2/3 area with moon-shaped terraces for horticulture	Bench terraces
Av. Losses/Year			
Runoff (mm)	114**	57	95
Runoff (%)	7.0	3.5	5.8
Soil (tonnes/ha)	40.9	3.0	2.3
Organic carbon (kg/ha)	698	40	34
Av. P_2O_5 (kg/ha)	0.15	0.02	0.01
Av. K_2O (kg/ha)	7.1	0.5	0.5

Source: Singh, 1981

**All the values are average for four years except for shifting cultivation which was fallowed after two years. The data for shifting cultivation are therefore for an average of two years.

and is often coconut-based. The home garden system on sloping lands is practised in Java. This system also needs to be explored for adaptation and adoption. Only one example of home gardens in the Mikir hills in the northeastern hills is reported (Maikhuri and Ramakrishnan, 1990), in which arecanut, betel leaf, banana and black pepper are grown together. This practice requires an annual input of Rs. 1,750 and provides an annual output of Rs. 15,352. This works out to a benefit : cost ratio of 8.8. Keeping of pigs and poultry is also suggested as a source of additional income for settled agriculture. However, all the above suggested alternatives take for granted that the people, whose social fabric and life are woven around shifting cultivation, will be willing to adopt settled agriculture. Ramakrishnan (1992) reports that no significant impact has yet been made by any model since the suggested innovations are in conflict with the societal organisation of the tribals. Further, settled agriculture cannot be easily adopted without addressing the question of landownership and distribution. In the northeastern hill region land is owned: (1) by the villagers collectively, (2) by the chiefs who distribute land among the individual households for shifting cultivation, and (3) by the individual families (Ganguly, 1980). Therefore apart from the technical aspects, and economic and ecological viability of the model, it would be necessary to address the land tenure and ownership aspects on the one hand, and mount an intensive training and demonstration effort on the other hand. Settled agriculture could be adopted on its own merit due to population pressure. Even in that case, however, the issues mentioned above will need to be addressed and resolved.

Another innovation which has been developed in the Philippines for weaning away shifting cultivators from shifting cultivation or cultivating on hillsides is sloping agricultural land technology (SALT). SALT is a modified version of alley cropping. The concept of SALT was initially articulated by Benge in 1976 (quoted by Caleda and Esteban, 1981), when he suggested that *Leucaena leucocephala* may be planted in five-metre wide bands (each band having five rows, one metre apart) alternating with 10 m bands of cultivated land to raise crops of maize, paddy, cassava, pineapple and other cash crops. The SALT technology involving *Leucaena* and *Gliricidia* as hedge rows was then developed in the early eighties by the Mindanao Baptist Rural Life Centre in the Philippines. By 1985, SALT had received wide publicity and acceptance (Watson and Laquihan, 1985). Considering the variable socio-economic and edaphic conditions, two more versions of SALT have evolved. SALT-II integerates goat production by cultivating agricultural and fodder crops. It is simillar to SALT-I except that SALT-II provides space for an alley of agricultural crops on one side of the hedge row, and an alley of fodder crops on the other side. The ratio of agricultural crops, trees and fodder is 40:20:40. SALT-III is small-scale reforestation integrated with food production, in the ratio of 60% trees and 40% food crops.

Due to very wide publicity, endeavours are made to introduce SALT in India also—more specifically in Orissa (Social Forestry Project) and the Himalaya. However, neither are the technological aspects clearly developed nor are the economic aspects clearly worked out. This practice therefore needs to be researched with respect to its technological components, economics and ecological impacts before it is adopted for development action in India.

5.2 TAUNGYA

Taungya is a silvoagriculture system of establishing departmental forest plantations in which agricultural crops are grown on a temporary basis in between regularly arranged rows of a forest tree species. This practice has been adopted by foresters with the main goal of reducing the formation, and maintenance costs of forest plantations, on forest lands under possession of the forest departments. *Taungya* is a Burmese word, but it is believed that the system is much older, having originated in China. It was introduced in India by Brandis in 1856. The first *taungya* plantations were raised in 1863 in north Bengal, in 1886 in Sylhet, and in 1890 in Coorg (Seth, 1981). It was introduced in Orissa in 1900 (Das and Mahapatra, 1984). However, regular *taungya* cultivation was not taken up until 1911 when it was used for raising *Shorea robusta* plantations in north Bengal, where it was soon extended to *Tectona grandis* in 1912. It was not until 1923 that it was adopted to regenerate failed sal (*Shorea robusta*) coppice coups in UP (Seth, 1981). In south India the system is referred to as *Kumri* cultivation.

Distribution

Taungya is no longer practised extensively in India. It is at present more of a historical interest. *Taungya* was practised extensively in Kerala, West Bengal, Uttar Pradesh; to a little extent in Tamil Nadu, Andhra Pradesh, Orissa, Karnataka, Maharashtra and the northeastern hill region. For the establishment of forests under rainfed conditions, it was practised in areas with assured annual rainfall of over 1,200-1,500 mm. Total area under *taungya* was very small, not exceeding 10,000 ha per annum. The owner of the land and hence of the trees was the state forest department.

The Practice

Traditional *taungya* consists of land preparation for tree plantation, growing agricultural crops for one to three years after tree plantation and then moving on to another area to repeat the cycle. In some cases crops may be grown one year before the plantation is undertaken.

Components

A large variety of crops and trees, depending on the soil and climatic conditions, and the state concerned were grown (Table 29). In West Bengal, usually a mixture of trees was grown. These mixtures varied according to the agroecological region as follows:

Plains

Shorea robusta, Michelia champaca, Tectona grandis, Chukrasia velutina, Lagerstroemia speciosa, Toona ciliata, Gmelina arborea, Bombax ceiba, Dalbergia sissoo, Acacia catechu, Anthocephalus chinensis, and *Bischofia javanica*

Foot hills

(up to 600 m): *Terminalia mayriocarpa, Tectona grandis, Chukrasia velutina, Shorea robusta, Acrocarpus fraxinifolius, Duabanga grandiflora, Betula cylindrostachys, Bombax ceiba* and *Schima wallichii*

Middle hills

(600-1,800 m): *Symingtonia populanea, Betula alnoides, Michelia lanuginosa, Cuppressus cashmeriana, Machilus edulis* and *Alnus nepalenis*

Upper hills

(1,800-2,400 m): *Cryptomeria japonica, Tsuga dumosa, Acer campbellii, Quercus* spp., *Juglans regia* and *Pinus patula.*

Some important practices are described here.

Taungya with Teak

The establishment of a teak plantation in Kerala was the most remunerative *taungya* practice for the forest department as well as the farmers. After clear felling of the forest, the forest department leased out the land to farmers. Due to land hunger as well as the possibility of growing highly remunerative crops (like paddy, tapioca, ginger, and turmeric), there

Table 29: Crops grown under the *taungya* practice

State	Tree crop	Associated agricultural crops and yield (kg/ha)
Uttar Pradesh (UP)	*Shorea robusta, Tectona grandis, Acacia catechu, Dalbergia sissoo, Eucalyptus* spp., *Poplar*	Maize (946), paddy (862), sorghum (1062), pigeon pea (1204), soybean, wheat (380), barley (1019), chickpea (447), rape seed (392), and miscellaneous
Andhra Pradesh (AP)	*Cashew* (*Anacardium occidentale*), *Tectona grandis, Bombax ceiba, Bamboo, Eucalyptus* spp.	Hill paddy (3017), Sweet potato (168); pigeon pea (180) + Millets (1160), groundnut
Karnataka (Coorg)	*Sandal* (*Santalum album*) with *Cassia siamea*	–
Kerala	*Tectona grandis, Bombax ceiba, Eucalyptus* spp.	Paddy (2000 to 3000), tapioca (14000 to 20000), ginger (12000), turmeric (12000)
Assam	*Shorea robusta*	Paddy
Tamil Nadu	*Tactona grandis, Bamboo, Eucalyptus* spp., *Santalum album, Tamarindus indica, Acacia nilotica, Acacia mearsnii, Ceiba pentandra,* cashew, rubber	Millets, pulses, sesamum, groundnut, cotton, tapioca, potato, medicinal plants like *Catharanthus roseus*
West Bengal* (WB)	*Tectona grandis, Shorea robusta, Schima wallichii, Cryptomeria japonica, Quercus* spp., *Michelia doltsopa*	Paddy (1800), maize, millets, ladyfinger, pineapple, hemp, turmeric (1200 to 2000), ginger (1500), potato, mustard (200), sesamum, cotton (500), buck wheat, pulses
Maharashtra	*Tectona grandis*	Sunnhemp, jute, mesta (1500 air dry), sunflower, castor etc.
Bihar**	*Tectona grandis, Dalbergia sissoo*	Maize, finger millet, paddy, sesamum, groundnut, soybean
Andaman and Nicobar Islands	*Pterocarpus* spp.	Sugar cane, maize

Source: Seth, 1981; *Lahiri, 1984b (yield data); **Mishra and Prasad, 1980.

was a great demand for land and the leases had to be auctioned. The lease value of land is reported to have varied from Rs. 650 to 1,063 per hectare during the period 1974-1976 (Nair, 1981). The average value being Rs. 1,000/ha. The lease rent offered by the *taungyadar* (*taungya* cultivator) depended on the location and convenience of the cultivator as well as the food sufficiency/security condition at that time. The forest department planted the tree crop and then gave possession of the land to the *taungya* lessee. The crops grown were paddy and tapioca in one practice and turmeric and ginger in another practice. Profits fluctuated but on an average a *taungyadar* made a net profit of about Rs. 2,000/ha/annum when paddy and tapioca were grown. In the case of ginger and turmeric, a net profit of Rs. 3,500/ha/annum was obtained (Nair, 1981). In Meghalaya, agricultural crops were taken for one year in teak plantations (Suchiang, 1984).

Taungya with Eucalyptus and Wattle

Establishment of *Eucalyptus globulus, Acacia mearsnii* (wattle) and pines in the high hills in Tamil Nadu was achieved by allowing cultivation of potato, peas, cauliflower, cabbage and radish in between the tree rows. In the drier regions millets, groundnut and cotton were grown as agricultural crops. Sometimes a medicinal plant like *Vinca rosea* was cultivated.

Taungya with Cashew

Establishment of cashew plantations in the coastal belt of Andhra Pradesh is achieved by allowing the cultivation of groundnut in between the tree rows (Reddy, 1981). This is a recent practice, less than three decades old. The lease is auctioned and the *taungya* lessee is allowed to grow groundnut. The lease rent varies from Rs. 500 to 1,000/ha. Since the east coast of India receives both the southwest and northeast monsoons, it is possible to grow two crops of groundnut per year. It is a condition of the lease that groundnut leaves are not to be removed from the land (Reddy, 1981). This practice has now been extended into the interior districts of Andhra Pradesh (Khamam and Adilabad) on red loam soils. It is also practised in south Arcot, Tiruchi, and Pudukottai districts of Tamil Nadu (Shanmugasundaram, 1984), where cashew is intercropped with groundnut, sesamum, pearl millet, finger millet and chillies.

Taungya with Poplars

Since 1962, big farmers have taken to *taungya* system to establish poplars in the Terai region of UP. Since this is practised under irrigated conditions on a large commercial scale and agricultural crops are taken for a longer period, the practice has already been described in section 4.4.

Taungya with other Crops

Turmeric and ginger have been successfully grown under the shade of *Dipterocarpus macrocarpa* in Arunachal Pradesh (Bhattee, 1984).

The forest department in West Bengal introduced medicinal plants (*Dioscorea, Costus, Vinca*), spices (turmeric, ginger, pepper), pineapple, ramie, citronella, bamboos, gladioli as intercrops in *taungya*. In general, the growth of agricultural crops was better in conjunction with slow-growing species like *Shorea robusta*, than with fast-growing species like *Tectona grandis*.

Resource Input and Utilisation

The owner of the land was the forest department. The management unit was 0.4 to 1.6 ha per *taungya* cultivator. Some classes of *taungya* cultivators are not accustomed to live entirely on dry crops and expect to supplement these with paddy, as for example in Assam and West Bengal. The area of paddy cultivation allotted is usually kept to the minimum for the real requirements (0.4 ha), otherwise there is a resultant neglect of *taungya*. Where cultivation continues for two to three years, it is sometimes customary to allot an additional new area every year. In case the land lease is auctioned, for example, for teak and *Eucalyptus globulus* as in south India, and the returns are high, the management unit may be 3-10 ha. Where there is land hunger and there is a keen demand for land, a rent of Rs. 2.5-5/ha may be collected by the forest department. Where the returns are high, the land lease values are Rs. 1000/ha in Kerala and Rs. 200-460/ha in Tamil Nadu. The forest department establishes the tree plantation and the cultivator provides labour to keep the plantation clean and free of weeds while cultivating the agricultural crops. Usually all inputs for the crops are provided by the cultivator. The forest department may provide rudimentary facilities for staying of the *taungyadars* close to the forest plantations. In the past, *taungya* cultivators had the obligation to provide labour for aligning, staking, collection of seedlings, planting and cutting bridle paths. Currently all these activities, perhaps with the exception of cutting bridle paths, are performed by the forest department. In some cases for example in West Bengal, the practice of making cash payment to the farmers at the end of a specified period is prevalent. It is usually Re. 1/100 plants or Rs. 25-30/ha with a sliding scale, penalising poor work.

Protective and Production Aspects

If the crop receives fertilisers, for example in the case of potato, paddy, tapioca, ginger and turmeric, the tree crop benefits considerably. Soil working and weeding done for agricultural crops benefits the trees also. If agricultural cropping is prolonged, not only does the crop yield decrease, but in some cases it may be harmful to the trees also. For example, a definite loss of teak growth has been demonstrated in south India for each successive agricultural crop, including the first crop (Nair, 1981). In the case of cashew plantation leaves of the groundnut crop, which are required to be left on the ground, benefit the trees, much more so when the inherent fertility of the coastal sand is low (Reddy, 1981). The cashew plant itself is recognised to be capable of binding coastal sand, preventing its drift and dune formation due to strong winds. When forest trees are grown on sloping or steep land, a certain amount of soil erosion occurs due to the growing of agricultural crops, even more so when the *taungya* cultivator cultivates the land up and down the slope. Soil fertility and conservation are not his concern; however, it is felt that the damage done by the cultivator in two to three years of cropping is compensated for by the next 40-80 years of forest growth. The crop yields obtained under *taungya* from different states in India are reported in table 29.

According to Sowani (1980), mesta (*Hibiscus sabdariffa* and *H. cannabinus*) seemed to benefit the growth of teak in Maharashtra. Ginger is reported to do better than tapioca in *Eucalyptus* plantations in Kerala (Kutty, 1980).

Since the results of tree plantations by *taungya* are usually good, the output from the forest department's viewpoint is very satisfactory. However, actual yields are not reported as *taungya* plantations are also treated as part of a normal forest department plantation programme and normal yield tables are applied to them. In the net analysis, the cultivator gets very little as compared to the total gain by the forest department (Seth, 1981). The greatest possible benefit of the *taungya* system is that it could help to change the destructive shifting cultivation in northeast India into a successful means of regeneration of forests. However this will require imagination and a great concern for the well being and welfare of shifting cultivators, and their social and cultural life.

Recently Kaul and Kapoor (1992) have reported a *taungya* model (*sic*) on community wasteland for the semi-arid Arravalli hills in Rajasthan. In this model ridges were made at 3 m spacing across the slope, bamboo (*Dendrocalamus strictus*) rhizomes were planted at 3 m spacing on the upper side of the ridges in 60 × 60 × 45 cm pits, and castor (*Ricinus communis*) was dibbled at 1 m spacing on both sides of the ridge, in 9 ha of land. Bamboo plants were hand-watered seven times during the first two years. Over a period of six years, this model produced 7.5 tonnes of natural grass (during the first three years); 5.35 tonnes of castor seed (during the first two years); 17,450 numbers of bamboos 2 to 2.2 m long and 2.5 to 3 cm diameter; 33,050 numbers of bamboos 2 to 2.5 m long and less than 2.5 cm diameter; 36.4 tonnes of bamboo waste including branches, and 2,000 number of entire culms, during the fifth and sixth years after planting. Since the canopy closed during the fourth year after planting of bamboo, no grass yield was obtained after the third year, and for the same reasons no castor seed was obtained in the third year. As against a total expenditure of Rs. 95,120, an income of Rs. 3,24,950 was obtained at the end of six years. The model had a payback period of five years, b:c ratio of 2.81, 2.61 and 2.34 at 10, 14, 20% discount rate, respectively.

In India, attention was focussed on research efforts to identify the suitability of different agricultural crops for intercropping with forest species in relation to various soils, time of planting and harvesting, spacing, use of improved varieties etc., since the early twenties. Depending on the state-of-art in agriculture, forestry tree species involved, and the agro-ecological and socio-economic conditions, this matter has been reviewed from time to time by many authors, particularly in the twenties and thirties (Allan, 1916; Anonymous, 1921; Hay, 1924; Rowbotham, 1924; Trafford, 1924; Wood, 1924; Shaukat Hussain, 1925a, b, c; Glausson, 1927; Coffey, 1929; De, 1932; Shaukat Hussain 1933; Stewart, 1933; Howard, 1939; Manning, 1941; Griffiths, 1942). This information only has historical value at present, since in the recent times the emphasis on the tree species on the one hand, and management techniques of the agricultural crops and tree species on the other, have changed very radically.

Socio-economic Aspects

Taungya has both positive and negative aspects depending upon the situation.

Due to the prevailing acute land hunger, the practice provided an opportunity for a class of people to cultivate land, obtain employment and food. In some cases the practice proved to be very remunerative as has been already mentioned. Some farmers pass their whole life as *taungya* cultivators. They are experienced persons and know not only how to grow good agricultural crops but also forest crops. However, the practice is now considered to be highly exploitative of the poor who have to live the life of nomads, particularly when their income is very low (Seth, 1981). Due to recent awareness, *taungya* cultivators show a tendency of not handing the land back to the forest department. This leads to further encroachment on the forest lands (Shanmugasundaram, 1984). Even when the intention is not to grab the land, there is a tendency to neglect or even damage the tree crop to obtain a higher crop yield and/or extend the lease. The farmers are in some cases backed by local political forces. As a consequence, the forest departments are also wary of practising this system. For these reasons, this practice was discontinued in Orissa in as early as 1950 (Das and Mahapatra, 1984). At present, the practice is dead as far as the forest departments are concerned.

Improvement of the Practice

Since the forest departments have given up this practice, one could conclude that there is no further scope of work on *taungya*. However, there is nothing wrong with the practice as such, as can be observed from the success stories associated with the growing of poplars and *Eucalyptus* under irrigated conditions on farms. The same could be true, if the forest departments also pay attention to the problems associated with *taungya*. The practice as implemented by the forest departments had both infrastructural and socio-economic constraints which are out of tune with the modern times. With respect to the socio-economic conditions, the practice could still be viable and useful, provided a fair deal is given to the *taungya* cultivators by the forest departments. What was possible earlier, is no more practical today. With respect to the infrastructural constraints, there is a scope of mechanisation and commercialisation of the practice. Both the manual and mechanical systems can coexist as there is a very wide scope of work to be done.

There is a need for research on the compatibility of trees and agricultural crops, input requirements of crops and trees, economic analysis of large-scale mechanised operations and benefit : cost analysis. The system should be fair to the farmer as well as to the forest departments. However, in the case of the forest departments, the social, long term and on site/off site benefits such as soil conservation, availability of both surface and underground water, reduction in sediment and improvement in overall ecological balance, must also be taken into account.

5.3 OTHER SPECIFIC AGRICULTURAL USES ASSOCIATED WITH FORESTS

Forests are used for many other activities bearing on agriculture, if the term "agriculture" is used in its broadest sense—not only covering crops and animal husbandry, and horticulture but also apiculture, floriculture, pisciculture, growing of medicinal and aromatic plants, etc. This aspect was recognised when the classification of agroforestry systems was proposed (Tejwani, 1987a). More recently, Australian scientists have also emphasised the importance of "alternative products from trees and shrubs" in sustainable agriculture in Australia (Boutland et al., 1992). Tejwani and Lai (1992) while describing agroforestry profiles in the Asia-Pacific, have also focussed on non-wood forest products. In view of the production and extraction of non-wood forest products, and the fact that very large numbers of people—both forest dewellers and non-dwellers—depend on them, assumes significance as an important agroforestry practice. While non-wood forest products contribute considerably to the revenue of forest departments, the information available is not updated. The benefits which accrue to the people, directly as well as indirectly are also neither documented nor understood. Here, an attempt is made to briefly describe the types of different non-wood forest products and their importance as forest-based silvoagriculture practices.

Edible and Industrial Oil Seeds

India has nearly 100 tree species which bear oil seeds; of these, only 10 to 12 tree species have so far been tapped (Gupta and Guleria, 1982). Some of the edible oil-yielding plants in the eastern Himalaya are *Lindera neesiana, Diploknema butyracea, Rhus griffithii* (Singh and Pazo, 1981). *Shorea robusta,* which grows throughout the subtropical zone at the foothills of the Himalaya and Chotanagpur plateau, is also a source of edible oil. Some edible and nonedible oil-yielding tree species in the western Himalaya are *Shorea robusta, Azadirachta indica, Diploknema butyracea, Madhuca latifolia, Mallotus philippinensis, Melia azedarach, Sassuria lappa* (root), *Valeriana wallichii* (rhizome), *Michelia champaca,* and *Mimusops elengi* (Khosla, 1984). According to the National Commission on Agriculture (NCA, 1976) there is a potential of 6.6 million tonnes of oil seeds and one million tonnes of oil but only 233, 515 tonnes of seed and 59,000 tonnes of oil were extracted. Other oil seeds of industrial importance are *mahua (Madhuca latifolia), neem* (*Azadirachta indica*), *karanj* (*Pongama pinnata*), *kusum* (*Schleichera oleosa*), *khakan* (*Salvadora oleoides*), *kokam* (*Garcinia indica*), *nahor* (*Mesua ferra*), *undi* (*Calophyllum inophyllum*) and various other miscellaneous oil seeds of tree and shrub origin (NCA, 1976).

Rearing of Silk Moths

India produces all four types of silk: mulberry, *tussar, eri* and *muga*. Mulberry silk production is the maximum (about 82%), followed by *tussar* silk (about 12%), *eri* silk (about 5%) and *muga* silk (about 1%). The mulberry and *eri* silkworms are completely domesticated, and have to be fed and reared indoors. The *muga* silkworm is also semi-domesticated. Further, these three types of silkworms flourish on tender leaves, and therefore the trees have to be pollarded to promote profuse branching. In fact any such activity in the forest would cause damage to the trees.

The *tussar* silkworms, on the other hand are wild, and have to be left in the open to feed at will and grow. *Tussar* silkworms are found in almost all parts of the country, but relatively more abundantly in Bihar, Madhya Pradesh and Orissa. *Tussar* silkworms feed on leaves of *Terminalia tomentosa, T. arjuna, Shorea robusta, Lagerstroemia parviflora, Eugenia jambolana, Quercus leucotrichophora, Quercus* spp., *Anogeissus latifolia* and *Madhuca latifolia* etc. (Singh and Pazo, 1981; Lahiri, 1984a; Khosla, 1984; Gupta, 1991; Gupta and Guleria, 1982). There is a great potential for *tussar* silk production. This is a labour-intensive activity and can provide employment to the forest dwellers.

Food

As many as 800,000 people, who are still food gatherers and hunters, depend largely on forests for their food and survival (Tewari, 1986). Forests also provide a complementary source of human food such as fruits, leafy vegetables, flowers, succulent shoots, tubers, roots, seeds, honey, juices, gums and mushrooms. These are consumed in normal times, and much more during scarcity periods. A number of tree species provide edible products. More than 85 of these have been listed (Anonymous, 1972). In the western Himalaya the wild forest plants used as a source of food are *Bauhinia vahilii, Cyperus rotundus, Dioscorea belophylla, Moringa oleifera, Nelumbo nucifera* (rhizomes, roots and tubers); *Bauhinia variegata, Capparis decidua, Capparis spinosa,* and *Phoenix acaulis* (buds); *Cissus repanda, Phoenix sylvestris* and *Ehretia leavis* (sap and latex); *Ehretia leavis* (bark); *Dendrocalamus strictus;* and *Ficus lacor* (young shoots); *Agave americana; Chenopodium album, Ficus glomerata, Moringa oleifera* and *Tamarindus indica* (leaves); *Buchanania lanzan, Chenopodium album, Coix lacryma-jobi, Dendrocalamus strictus, Diploknema butyracea, Dictaria sanquinalis, Heritiera moinor, Fagopyrum* spp., *Jugulans regia, Lathyrus sativus, Mangifera indica, Nelumbo nucifera, Pinus girardiana, Shorea robusta, Tamarindus indica* and *Trapa bispinosa* (seeds) (Khosla, 1984). Some of the common edible mushrooms collected and consumed in the western Himalaya are *Morchella esculenta, M. crassipes, M. conica, M. hybrida, Halvella* spp., *Agricus campestris, Copernus atramentrivs, C. comatus, Lycoderdon* spp., *Calvaria* spp., *Lentinas* spp., *Lactarius* spp., *Cantharellus* spp., *Hydnum* spp. etc. Young fronds of ferns (for example *Diplazium polypodiodes*) are used as food by man (Khosla, 1984).

Other trees which provide edible fruits in many other parts of India are *Anacardium occidentale, Emblica officinalis, Aegle marmelos, Feronia elephantum, Artocarpus lokoocha, Syzygium cuminii, Annona squamosa, Carissa opaca* and *Morina deifera;* edible flowers of *Madhuca latifolia* and *M. longifolia*; edible root and tubers of *Amorphophalus companulatus, Discorea belophylla, D. oppositifolia, Imopoea aquatica* etc. (Gupta, 1991).

Some of the fruits are very valuable and considered as delicacies, for example, *Buchanania lanzan, Anacardium occidentale* and *Pinus gerardiana* In spite of such an important contribution of forests to edible products, very little attention has been paid to quantify the products or improve the management of trees and the extraction of food.

Flowers

There are 1,250 kinds of orchids in India, of which 600 are reported to occur in the northeastern hill region. There is a growing demand for some of these, both inside and outside the country (Singh and Pazo, 1981). Some of the orchid species growing is the western Himalaya, which are of ornamental or medicinal value, are *Coelogyne ovalis, C braeeox, Cymbidium cigantum, Cypripedium cordigerum, C. himalaicum, Dendrobium amoenum, D. chrysanthum, D. fimbriatum, Eulophia campanulata, E. compestris, E. hormusjii, Microstylis cylindrostachya, M. wallichii, Phajus albus, Rhynchostylis retusa* and *Vanda cristata* (Khosla, 1984).

Industrial Raw Materials

These include gums, resins, gum-resins, oleo-resins, grasses, bamboo, canes, essential oils, drugs, spices, poisons, insecticides, leaves, lac and its products (NCA, 1976).

Gums

The most important of all Indian gums is *karaya,* obtained from *Sterculia urens,* and *S. vilosa.* Other important gums are *ghatti* (from *Anogeissus latifolia); jhingan* (from *Lannea coromandelica*); *katira* (from *Cochlospermum religiosum*); Malabar *kino* (from *Pterocarpus marsupium*); Bengal *kino* (from *Butea monosperma*). *Acacia catechu, A. nilotica, A. modesta, A. senegal, Albizia chinensis, A. odoratissima, A. procera, Azadirachta indica, Bauhinia retusa, B. purpurea, B. racemosa, B. variegata, Feronia limonia, Mangifera indica, Bombax cieba, Terminalia alata* and *T. bellirica* also yield gum (Khosla, 1984).

Resins

The most important oleo-resin is that obtained from tapping the *chir* pine (*Pinus roxburghii*) and *kail* or blue pine (*P. wallichiana*). Resins are an important source of the revenue of the forest departments in Uttar Pradesh, Himachal Pradesh and Jammu and Kashmir. *Ferula asafoetida* and *Boswellia serratta* are also exploited for oleo-resins (Khosla, 1984). Other sources of oleo-resins are, *Dipterocarpus turbinatus* and *Kingiodendron pinnatum. Garcinia morella* and *Commiphora mukul* yield gum-resins. Hard resins are obtained from *Canarium strictum, Hoppea odorata, Shorea robusta and Vateria indica* (Gupta, 1991). Some work has been done on the technique of tapping of trees for gums and resins by the Forest Research Institute.

Grasses

Annually about 60,000-80,000 tonnes of *Sabai* grass (*Eulaliopsis binata* are collected and processed *in situ* and supplied to paper mills. Villagers collect and use it for rope making. Bamboos are used extensively for a very wide range of uses such as pulp, fuel, packing including basket making, furniture, ladders, mats, wood working, bamboo parquet and bamboo reinforced concrete. About two-third of bamboo produced is used in other ways than in industry (Purshotham, 1962).

Though there are over 100 species of bamboo which are found in India, *Bambusa arundinacea, B. tulda, B. polymorpha, Dendrocalamus strictus, D. hamiltonii, Melocanna baccifera* and *Ochlandra travancorica* are more important from the point of view of availability. However, *Dendrocalamus strictus* and *Bambusa arundinacea* are the two principal economic species, of which the former occupies the largest area and is the most common (Gupta, 1991).

Essential Oils

About 30 different types of essential oils are produced. The essential oil may be obtained from leaves and roots of grasses, and from leaves and wood of trees. The more important essential oils are lemon grass (*Cymbopogon flexuosus*); rosa or palmarosa (*C. martinii*); *Eucalyptus* oil (from *Eucalyptus globulus; E. citriodora); khus* or vetiver (from *Vetiveria zizanioides*); sandalwood oil (*Santalum album*); and linaloe (*Bursera penicillata*).

Medicinal Plants

Out of about 2,000 items of drugs recorded in the Indian Materia Medica, over 1,800 are of vegetable origin. About 128 of these are recognised in the Indian and British Pharmacoppea. Quite a few of them are obtained from forests. Drugs are derived from fruits, seeds, flowers, leaves, stems, roots of trees, shrubs, creepers and herbs. Solanaceous plants like *Atropa belladona, A. acuminata, Hyoscyomus niger* and *Datura* spp., which yield the propane group of alkaloids such as hyoscyamine, hyoscine, and metalodine are well distributed in the western Himalaya. *Atropa belladona* is cultivated in Kashmir at an altitude of 2,000 to 3,000 m; *Hyoscyomus niger* occurs through out the western Himalaya at an altitude of 1,500 to 3,000 m in the wastelands (Khosla, 1984). *Solanum khasianum, Costus* spp., *Dioscorea* spp., *Smilax* spp., *Gynocordea odorata* and *Strychnos nuxvomica* are found scattered in the forests of Manipur state (Sarma, 1984). *Banslochan,* a secretion found in the culms of bamboo, is used as a cooling tonic, aphrodisiac, and for the treatment of asthma and cough (Raizada and Chatterjee, 1956). A number of aromatic and medicinal plants are found in the alpine zone (for example *Aconitum heterophylus, Rheum emodi, Betula utilis, Carum bulbocastanum, Rhododendron companulatum,* and *Nardostachys jatamansi*), the temperate zone (for example, *Valeriana wallichii, Selinium tenuifolium, Morchella esculenta, Berberis aristata, Picrorrhiza kurroa* and *Skimmia laureola*), and the Siwalik ranges (for example *Terminalia chebula, T. bellirica, Emblica officinalis, Helicteres isora, Celastrus paniculatus, Acacia catechu, Withania somnifera, Mucuna prurita, Adhatoda vasica,* and *Pistacia integerrima*) of Himachal Pradesh (Sagwal and Gupta, 1984).

Due to continuous extraction, some of the species have become scarce in forests and efforts are being made to cultivate them. The important ones among these are *Dioscorea floribunda, D. composita*; two improved verieties of *Solanum khasianum* have been developed; *Datura stramonium, D. innoxia* and *D. metel* and their improved varieties; *Costus speciosus, Atropa acuminata, Rauwolfia serpentina* and *Cassia angustifolia* (Gupta, 1991). In spite of the abundance and importance of the medicinal plants growing naturally in forests, very little or almost no efforts have been made to develop cultural and other management practices to cultivate medicinal plants in forests (Gupta, 1991).

Spices

A large number of spices are obtained from forests, such as *kalajira* (*Carum carvi*); *galanga* (*Alpinia galanga*); cinnamon (*Cinnamomum zeylianicum*); small cardamom (*Elettaria cardamomum*); large cardamom (*Ammomum subulatum); tejpat (Cinnamomum tamala*); and pepper (*Piper nigrum; P. longum*).

Leaves

The most important, from the commercial point of view, are the leaves of *tendu/kendu* (*Diospyros melanoxylon*) used as wrappers of tobacco to produce *bidi* (Indian cigarette). The annual production of *tendu* leaf in the country is estimated to exceed 300,000 tonnes. Madhya Pradesh, Orissa, Maharashtra and Andhra Pradesh account for about 85% production of *tendu* leaves. Leaves of *apta* or *ashitri* (*Bauhinia racemosa*) are also used as wrappers in the manufacture of *bidi*. Leaves of *Bauhinia vahlii* are used for making cups and plates; they are collected on a commerical scale in Madhya Pradesh, Orissa and Andhra Pradesh (Gupta, 1991).

Lac

Lac is a resinous protective secretion of the tiny lac insect (*Laccifer lacca*). Lac finds a wide variety of applications in plastics, electricals, adhesives, leather and wool finishing, printing ink, sealing wax and other industries. Lac cultivation is done by about three million tribal cultivators. It is cultivated in Bihar, West Bengal, Madhya Pradesh, Uttar Pradesh, Orissa and Maharashtra. Its average annual production in 1966-1971 was about 30,000 tonnes (NCA, 1976), while it has a potential of about 50,000 tonnes annually. Two strains of the lac insect are recognised in India—the *Rengeeni* and the *Kusumi*, depending upon the host plant on which they feed. The former is raised on several host plants, the most important of which are *Butea monosperma* and *Zizyphus mauritiana*; the latter is raised on *Scheleichera oleosa* (Khosla, 1984).

Fibres

Agave sislana has a great potential if planted along fire lines, plantation boundaries, clearances under power transmission lines, and nursery and farm boundaries. As against the current annual production of about 2,500 tonnes of sisal fibre the total estimated annual requirement is 22,400 tonnes (NCA, 1976). The most common floss in India is that of *semul* (*Bombax ceiba*). The flosses from fruits of *kapok* (*Ceiba pentandra*, and *Eriodendron anfractuosum*), are elastic, buoyant and resistant to water absorption. They are therefore used for making life belts. India produces about 3,000 tonnes of *kapok* floss annually. *Kapok* trees are at present planted in Kerala and adjoining tracts. The cultivation could be encouraged in high rainfall areas to make the collection of *kapok* floss economical. Canal banks, road sides, edges of plantations, farm boundaries (see section 2.1.4) lend themselves for *kapok* plantations.

In the western Himalaya fibres are extracted from: *Grewia optiva, Bauhinia vahlii, Butea monosperma, Ougeinia oojeinensis, Marsdenia tenaissima, Ficus religiosa, F. bengalensis, Girardiana heterophylla, Kydia calycina, Cordia myxa, Albizia odoratissima, Erythrina suberosa, Moringa oleifera* and *Urena* spp. Silky flosses are obtained from *Bombax ceiba, Cochlospermum religiosum, Salix dephnoides, Populus ciliata, Writchtia tomentosa* and *Calotropis gigantea* (Khosla, 1984).

Based on the general use, the fibres are classifed as textile fibres, brush fibres, plaiting and weaving fibres, filling fibres, and fibres for paper making. The Indian cottage industries use the following more important fibres obtained from the forests: *Agave sisalana, Abroma augusta, Abutilon* spp., *Ananas comosus, Antiaris toxicaria, Boehmeria nivea, Borassus flabellifer Cannabis sativa, Cordia dichotoma, C. rottii, Girardinia heterophylla, Grewia elastica, G. glabra, G. optiva, Hibiscus* spp., *Malachra capitata, Marsdenia tenacissima, M. volubilis, Phormium tenax, Sensivieria roxburghiana, Sesbania bispinosa, Sida rhombifolia, Sterculia foetida. S. urens, S. villosa, Themada_arundinacea, Trema orientalis, Typha elephantina, Urena lobata,* and *Oveocnide integrifolia.*

Tans and Dyes

These are produced from a very large variety of trees and shrubs. The chief vegetable tan materials are myrobalan (myrobolan) nuts and barks of *Aacacia mearsnii, A. decurrens, Cassia auriculata,* and *Acacia nilotica.* Apart from these, a number of other tanning materials are used by many cottage and small-scale tanneries all over the country, for example, *Anogeissus latifolia* leaves and bark; *karada* (*Cleistanthus collinus*) bark; *ghatber or ghont* (*Zizyphus xylopyrus*) fruits; *amaltas* or *sunari* (*Cassia fistula*) bark; *sal* (*Shorea robusta*) bark; *sain* or *saij* (*Terminalia alata, T. tomentosa*), and *arjuna* (*T. arjuna*) bark; *amla* (*Emblica officinalis*) leaves and fruits; *karonda* (*Carissa spinarum*). Bark of *Acacia* spp., *Alnus* spp., *Myrica sapida, Terminalia tomentosa*; flowers and fruits of *Mallotus philippensis, Butea monosperma, Toona ciliata,* and *Rubia cordifolia*; leaves of *Indigofera tinctora* and *Lawsonia inermis* yield vegetable dyes (Khosla, 1984).

Fruits of three different species are known in the trade as myrobalans—*harra* (*Terminalia chebula*), *bahera* (*T. bellirica*), and *amla* (*Emblica officinalis*). *Harra* is the most important of all; its supplies come mostly from Madhya

Pradesh and smaller quantities from Orissa, Maharashtra, Bihar, Andhra Pradesh, Tamil Nadu, and Karnataka. Wattle (*Acacia mearsnii*) bark is obtained from the Nilgiri and Palini hills of Tamil Nadu. The total annual requirement of vegetable tanning materials by India's leather industry is estimated to be about 250,000 tonnes.

Cutch

It is a dark-brown product rich in catechin tannic acid, and is obtained from the heartwood of *Acacia catechu*. *Cutch* has preservative properties and is utilised for dyeing canvas, fishing nets, mail bags and sail cloth, and is used in drilling operations. Another product obtained from *Acacia catechu* heartwood in the same process is *katha*, a comparatively pale product rich in catechin. It is used as an ingredient in preparing *paan* (betel leaf) for chewing. The annual production of *cutch* and *katha* is 7,000-9,000 tonnes from Andhra Pradesh, Gujarat, Madhya Pradesh, Maharashtra, Rajathan, Uttar Pradesh and West Bengal.

Dyes

The more important plant dyes are: (1) wood dyes (from *Acacia catechu, Artocarpus heterophyllus, A. lakoocha, Pterocarpus santalinus*, and *Caesalpinia sappani*); (2) bark dyes (from *Terminalia tomentosa, Acacia concinna, A. farnesiana, A. leucophloea, Alnus* spp., *Casuarina equisetifolia, Manilkara littoralis, Myrica esculenta,* and *Ventilago maderaspatana*); (3) flower and fruit dyes from *Mallotus philippensis, Woodfordia floribunda, Bixa orellana, Butea monosperma, Toona ciliata, Nyctanthes arbortristis; Mammea longifolia, Wrightia tinctoria,* and *Carocus sativus*; (4) root dyes from *Berberis aristata, Datioca cannabina, Morinda tinctoria, Punica granatum* and *Rubia cordifolia;* and (5) leaf dyes from *Indigofera tinctoria* and *Lawsonia inermis* (Gupta, 1991). Vegetable dyes have not been able to compete with the artifical dyestuffs, yet they have a speciality value in cottage industries.

Fisheries in Association with Forests

Mangroves and such other habitats yield fish for human food. None of these systems is even studied and described.

Improvement of the Practice

All the above-mentioned products in the language of foresters, unfortunately, are referred to as minor forest products. Minor forest products include all items of forest produce other than timber, smallwood and fuelwood. The term "minor" is used in relation to the major and traditional preoccupation of the foresters with timber and industrial wood. If one were to take into account the economic and social value of these so-called minor forest products one would refrain from using this term. Of late, realising the economic and social importance of these products the so-called minor forest products are being referred to as non-wood forest products. The National Commission on Agriculture (NCA, 1976) reported that in 1967-1970 the minor forest products contributed 30% of the total revenue of the forest department. It is reported that minor forest products have assumed greater commercial importance over the years. In 1980-81 revenue from the minor forest products was valued at Rs. 1,701 million as compared to Rs. 360 million in 1969-70; in terms of percentage value of total forest revenue, the minor forest products were 36% in 1980-1981 as compared to 28.9% in 1969-1970 (Tewari, 1986). Further, it is estimated that 60% of the total products are consumed locally and are not accounted for in the calculations of the revenue from forests (Gupta, 1991). Non-wood forest products play a vital role in the import-export trade of India. For example *tendu* leaves, gums, lac, essential oils, and medicinal plants are exported, and edible oils and tannins are imported. Very large numbers of tribal people depend directly on these products not only for their own use but also for their cash income. A study conducted in Bastar District of Madhya Pradesh showed that in an average household (with two adults, at least one child and one old person) as against the total average annual income of Rs. 1,750 the average income was Rs. 1,500 a year from the sale of minor forest products without any input or risk (Tewari, 1986). A similar study conducted by the Tribal Research Institute, Ahmedabad showed that 35% of the total earning of tribals in the Panchmahal district of Gujarat was from non-wood forest products.

It is reported that in the six tribal district of Orissa, as many as 60 million mandays are spent on the collection of non-wood forest produce, providing 25 working days per worker per year. Non-wood forest products are collected in Orissa for the Government Department/Agency (for example, *tendu* leaf and *sal* seed); for individual lessees (for example, arrow root, wax, honey, resin), and for cooperative societies (CDS, 1990). In the case of the Government Department and lessees, the tribals work as wage earners. It is reported that of the employment of 2.3 million man-years generated by the forestry sector, as many as 1.6 million man-years of employment is provided by these "minor" forest products.

Apart from the employment generated by gathering the non-wood forest products, there is a considerable scope for the generation of additional employment by small-scale processing enterprises. These include basket- and mat-making, broom-making, silkworm rearing, bamboo and cane furniture making, and handicrafts. Very little socio-economic information is available regarding the labour requirements, employment and income generation for indivduals/families. It is in the realm of conjecture that perhaps non-wood forest products provide more benefits to the local/individual/family economies than organised forestry operations. While the importance of the minor forest products in the daily lives of the millions of tribals, and other people, and in import-export trade is realised (NCA, 1976; PC, 1981a,b), yet it is unfortunate that very little significant work has been done to understand and improve the system biophysically as well as socio-economically. The National Commission on Agriculture has made a number of suggestions in 1976 for research and development activities. While attention is being paid to the management and administration of some important non-wood forest products such as *tendu* leaves, yet very little attention is being paid to research on increased production, better management, or improvement of varieties, for example, of various non-wood forest product trees. In a book on agroforestry and with our limited knowledge, it is perhaps sufficient to report a list of multipurpose tree species which provide these non-wood forest products along with fodder, and fuelwood (Appendix 2).

CHAPTER 6

PASTORALSILVICULTURE PRACTICES

Out of a reporting area of 305 m ha in India there are 67.1 m ha of forest land, 15.2 m ha of culturable wastelands and 11.8 m ha of permanent pastures and other grazing lands (MOA, 1992). All this area of 94.1 m ha, grows grasses, bushes and forest trees in varying proportions. Forest land is mostly owned and managed by the state governments, although some of it is owned and managed by villagers. Each village, or a group of villages, is allotted some land for grazing of cattle. Indian farmers have integrated crop farming with very large numbers of livestock. They supported 180 million cattle, 62 million buffaloes, 41 million sheep, 75 million goats, 1 million camel, 2 million horses, mules, donkeys, 8 million pigs, 0.13 million other livestock in 1977 (MOA, 1982). This livestock depends on grasses, and fodder from agricultural crops and trees for survival.

Most of the livestock graze freely in the village grazing lands and the adjoining forest lands where the villagers have the right to graze their animals. Livestock also graze at large in agricultural fields when no crops are growing. This is very well regulated by the traditional and social practices. The agro-ecological and edaphic conditions in India generally contribute to the growth of trees even in grasslands except in alpine lands; conversely, grass will also grow in forests, except perhaps in high rainfall tropical areas where the forest canopy is closed. The practices which have a major component of grass and grazing land and a scattering of trees are referred to as pastoralsilviculture practices in this book. These are practised extensively by farmers all over India; the types of grasses and trees varying with the regional agro-ecological conditions. Some of the practices are very well developed; for example, grassland and tree management by the farmers in the Deccan on their own land, and in the Rajasthan desert. In fact, farmers have generally developed traditional systems and know how to manage them.

6.1 GRASSLAND-TREE MANAGEMENT IN THE SEMI-ARID DECCAN PLATEAU

Distribution

This practice is prevalent in Andhra Pradesh, Karnataka, Maharashtra,

Tamil Nadu and Madhya Pradesh in the semi-arid tracts at an elevation of about 300 to 1,000 m amsl.

Climate

The climate is semi-arid with no winter. The coldest month is December during which a minimum temperature of about 12° C is recorded. The average annual rainfall is 500-1,300 mm in 45-70 rainy days occurring mainly during the southwest monsoon and some in the northwest monsoon.

The Practice

The grassland-tree combination system—locally called *kanchas* is very widely prevalent and supports animal husbandry in the semi-arid tracts. The practice is indigenous with natural grasses and trees. Planting of grasses and trees may also be done to obtain quicker results, and to grow desirable tree species like *neem* (*Azadirachta indica*) and tamarind (*Tamarindus indica*). The *kanchas* are owned by cultivators. Some are with the government forest departments.

Grasses

The lands with existing trees are left fallow and protected. Natural succession of the grasses takes place. The grasses are protected from grazing for one to three years depending on the economic condition of the farmer. The grasses fall mainly under the *Sehimia-Dicanthium* grassland type. A good stand of *Sehimia-Chrysopogon-Eremopogon-Heteropogon* is obtained on red soils, and *Dicanthium-Bothriochloa* is obtained on black soils. In gravelly and skeletal soils *Dicanthium-Cenchrus* grassland is found. These grasses grow fast and depending on the location and site conditions, reach the climax stage in four to 10 years.

Trees

Trees are representative of arid and semi-arid climates—*Acacia nilotica; A. leucophloea; A. planifrons; Madhuca latifolia; Azadirachta indica; Tamarindus indica; Melia azedarach; Annona squamosa; Tectona grandis; Hardwickia binata; Mundulea suberosa; Dalbergia sissoo,* etc. The commonly planted exotics are *Eucalyptus, Casuarina equisetifolia* and *Leucaena leucocephala*. Palms like *Borassus flabellifer* and *Phoenix dactylifera* are also common.

Trees are planted, or the existing trees growing naturally as per the prevailing agro-ecological and edaphic conditions are encouraged to grow, to the liking of the farmer. The trees are lopped every year for fodder. Custard apple (*Anonus squamosa*), mango, *Zizyphus* fruits, and tamarind are collected for home consumption and sale; *Mahua* (*Madhuca latifolia*) fruits yield edible oil, and its flowers are used for brewing alcoholic drinks; *neem* (*Azadirachta indica*) fruits are collected for sale.

Animals

Livestock are an integral part of the *kanchas*. The grasslands are grazed by sheep, goat, cattle and buffaloes. The animals provide milk, wool, leather, bone, and meat; they plough the land, and are used for transport and irrigation operations.

Yields

Resource input is restricted to closing the area, planting (wherever undertaken), and managing the grasslands and trees. In red soils the yield is 3 tonnes dry grass/ha/year in well-protected *kanchas*, otherwise it is 1.0 to 1.5 tonnes. Grass legume ratio is 8 : 1 to 44 : 1, and the key forage species may vary from 22 to 72%. In gravelly soils the yield is low, 0.5 to 1.0 tonne dry grass/ha/year. In black soils, grass yield varies from 1.5 to 10 tonnes green grass/ha/year, with grass : legume ratio of 2 : 1 to 8 : 1, and with 25 to 75% key forage species. The yield of grassland depends on the stage of succession and management. In a well-protected natural grassland in the black soil region the yield of green grass may be 1.5, 1.5-5, 5-10 and above 10 tonnes in the first, second, third and fourth year after closure, respectively. The stage of grassland could progress from *Aristida*, to *Eremopogon*, to *Bothriochloa* and *Dicanthium* in three to four years. A cultivated grassland of *Cenchrus ciliaris* may yield 1-2, 5-10, 10-12 and 12-15 tonnes of green forage/ha in the first, second, third and fourth year, respectively. It takes about seven years of rigid protection to obtain the *Sehimia* subtype grassland; the *Dicanthium* subtype grassland can be achieved with four years of rigid protection. Since the farmer generally provides for only one to two years of protection, he does not achieve these subtypes. Consequently he gets only 20-50% of the anticipated climax yield. *Acacia nilotica* in the black soil *Dicanthium* subtype grassland may produce 14.0 to 55 m^3 of fuelwood in a period of 12 to 30 years depending upon the site quality and soil depth. In the case of trees also, now shorter rotation trees are preferred, for example, *Casuarina equisetifolia, Melia azedarach, Eucalyptus* spp. and *Leucaena leucocephala. Azadirachta indica* fares better in all the locations. *Madhuca latifolia* does better in red soils.

Protective and Production Aspects

The establishment of *kanchas* is an ideal land-use practice, since the land is under grass-cum-tree cover. If deferred cutting of grass is practised, the system is ideal for conservation, since in the case of continuous grazing, there is a tendency to overgraze the land. No data on the improvement of soil fertility or productivity are available.

The price for fodder and fuelwood is determined by the proximity of the *kanchas* to towns and cities, and transport facilities. In semi-arid areas the agricultural crop may fail once, and fall below the normal in two years out of five. The grassland-tree combination gives a stable and certain income to the farmers. A farmer may get as much as Rs. 900-1,800/ha by the sale of grass as compared to Rs. 575-1,625 by the growing of crops. The grass yield is certain; the crop return is uncertain and is accompanied by erosion hazards. This practice has degraded and degenerated a lot during the recent times due to an increase of pressure on the land and also as a result of inadequate maintenance.

Improvement of the Practice

No research on *kanchas* has been carried out. It would appear that the general principles of maintaining and managing the grasslands and trees would be applicable to the biophysical components of the practice. However the socio-economic aspects of the practice are the least understood.

6.2 GRASSLAND-TREE MANAGEMENT IN THE ARID ZONE

The importance of *Prosopis cineraria* and *Zizyphus nummularia* in the agrosilviculture system in the hot arid zone has already been discussed in section 2.1.1. These two species play a vital role in the pastoralsilviculture practice also. The community and village grazing lands in Rajasthan are called *Oran* or *Bir.* Both these species are encouraged and maintained together in the *orans.* The general distribution of both these species has been discussed earlier under the agrosilviculture practices (section 2.1.1). Village common grazing lands with shallow soils and *kankar* pan at 30-45 cm soil depth are generally dominated by *Zizyphus nummularia* shrubs. There are some big *Zizyphus orans* dedicated to deities in Bikaner, Jaisalmer and Jodhpur districts of Rajasthan. These *orans* have 75-90% dominance of *Zizyphus* plants. The rest 10-25% is generally shared by *Prosopis cineraria, Balanites aegyptiaca,* and *Capparis decidua.* Here the shrub density with clear bole and crown ranges from 190 to 120 plants per hectare (Saxena, 1981).

While the role of *Prosopis cineraria* and *Zizyphus nummularia* in the arid zone grassland *oran* management is well recognised, yet there are a large number of silvopastoral practices prevalent in the area. These practices have been classified according to physiographic, soil, and rainfall conditions (Bhati, 1984). The combination of the tree and grass species under various physiographic and soil conditions is as follows:

A. Sand dunes and sandy plains

Grasses: *Lasiurus indicus, Panicum antidotale, Saccharum bengalensis, Cenchrus ciliaris.*

Trees: *Acacia tortilis, Prosopis cineraria, P. juliflora, Albizia lebbek, Acacia senegal, Dichrostachys glomerata, Cassia siamea, Tamarix articulata, Holoptelia integrifolia, Colophospopermum mopane, Calligonum polygonoides.*

B. Sandy loam and sandy clay loam

Grasses: *Dichanthium annulatum, Heteropogon contortus.*

Trees: *Acacia tortilis, Prosopis juliflora, Albizia lebbek, Zizyphus nummularia, Holoptelia integrifolia, Dalbergia sissoo.*

C. Sandstone and rocky sites

Grasses: *Sehemia nervosum, Dichanthium annulatum.*

Trees: *Anogeissus pendula, Acacia senegal, Capparis decidua, Zizyphus nummularia, Acacia tortilis, Butea monosperma.*

D. Well drained sandy alluvial soils

Grasses: *Cenchrus ciliaris, C. setigerus.*

Trees: *Acacia tortilis, Dichrostachys glomerata, Ailanthus excelsa, Azadirachta indica, Albizia lebbek.*

E. Low lying saline soils

Grasses: *Sporobolus marginatus, Chloris virgata.*

Trees: *Prosopis juliflora, Tamarix articulata, Salvadora oleoides, S. persica.*

The classification on the basis of rainfall is as follows:

1) Rainfall zone 150 to 300 mm

Grasses: *Lasiurus sindicus.*

Trees: *Prosopis juliflora, Acacia tortilis, A. senegal, Dichrostachys glomerata, Calligonum polygonoides, Zizyphus nummularia.*

2) Rainfall zone 300 to 400 mm

Grasses: *Lasiurus sindicus, Cenchrus ciliaris, C. setigerus, Panicum antidotale.*

Trees: *Acacia tortilis, A. senegal, Prosopis juliflora, P. cineraria, Tecomella undulata, Colophospopermum mopane, Zizyphus* spp., *Calligonum polygonoides, Ricinus communis.*

3) Rainfall zone above 400 mm

Grasses: *Cenchrus ciliaris, C. setigerus, Panicum antidotale, Dichanthium annulatum, Sehima nervosum, Saccharum bengalensis.*

Trees: *Prosopis cineraria, Acacia tortilis, A. senegal, A. nilotica* sub. sp. *indica, Tecomella undulata, Ailanthus excelsa, Albizia lebbek, Dalbergia sissoo, Dichrostachys glomerata, Ricinus communis, Zizyphus* spp., *Cassia auriculata.*

Studies on the scientific management of the arid zone grasslands have been in progress for a long time at the Central Arid Zone Research Institute, Jodhpur. Studies on tree density and yield of grass have indicated that under arid climatic conditions, the yield of *Cenchrus ciliaris* dry grass was rather low (812 kg/ha with grass only, and 656 to 980 kg/ha under tree species). Growing of trees (*Acacia tortilis, Colophospopermum mopane* and *Hardwickia binata*) did not adversely affect the yield for the two years of experiment. *Acacia tortilis* gave more yield of grass at 5 × 5 m spacing, while the other two tree species gave better yield at 10 × 10 m spacing (Bhati, 1984), The relative returns from the tree component are not available. Ahuja (1980) reported yield of grasses under four tree species planted in 1959 at different spacings. The grass yield was recorded, when the trees were 14 to 18 years old, in 1972 to 1976. The data for six-year forage production (Table 30) indicated that the yield of forage (1.1 to 1.5 tonnes/ ha) was the highest under *Prosopis cineraria* among the four tree species evaluated, whereas it was the lowest under *Acacia senegal* (0.6-0.7 tonnes/

ha). It was also concluded that the proportion of perennial species contributing to forage production was the highest under *P. cineraria*.

Prosopis cineraria tree when cut above the ground level, produces numerous new buds which give rise to several (5-12) new shoots. These in a year's time assume a bushy structure. In an *oran* the coppiced shoots are continuously grazed leading to a "cushion" form of the crown. Under severe grazing and trampling stress, the new sprouted branches spread horizontally and provide feed to small ruminants (Saxena, 1984).

Table 30: Average forage yield (t/ha) under different tree species

Experiment no. & period of observation#	Particulars	Tree species			
		Prosopis cineraria	*Tecomella undulata*	*Albizia lebbek*	*Acacia senegal*
Expt. 1 1972-76	1. Total forage production	*1.5	*1.5	*1.4	0.6
	2. Contribution of perennial spp.	1.1	*0.8	*0.9	0.3
	3. Perennial grass species %	74.8	53.7	57.9	52.5
	4. Trees/ha	566.6	383.3	283.3	1516.6
Expt. 2 1972-75	1. Total forage production	*1.5	*1.5	*1.4	0.7
	2. Contribution of perennial spp.	*1.2	0.8	0.9	0.4
	3. Perennial grass species %	78.3	51.1	66.5	58.8
	4. Trees/ha	*1377.5	*942.5	*688.8	3724.4
Expt. 3 1972-76	1. Total forage production	*1.2	–	*1.1	0.7
	2. Contribution of perennial spp.	0.9	–	*1.0	0.6
	3. Perennial grass species %	74.7	–	85.3	82.4
	4. Trees/ha	*1743.3	298.0	298.3	*1812.4
Expt. 4 1972-77	1. Total forage production	*1.1	*1.1	0.9	0.7
	2. Contribution of perennial spp.	*1.0	*1.1	0.9	0.7
	3. Perennial grass species %	97.9	99.2	97.1	98.8
	4. Trees/ha	*1264.8	1111.1	362.0	*1280.5

Source: Ahuja, 1980

#Experimental details: Expt.1: two planting methods of trees, at 7 × 7 m spacing; Expt.2: five soil working methods, and two methods of sowing trees at 12 × 10 m spacing; Expt.3: three ages of seedlings, at 12 × 10 m spacing; Expt.4: three times of planting, at 12 × 9 m spacing

*: Values in a horizontal line do not differ significantly

Shrubs of *Zizyphus nummularia* also have remarkable powers of regeneration through root suckers. They start sprouting by February-March. The tender branches with leaves are browsed by goats.

Prosopis cineraria and *Zizyphus nummularia* plants in an *oran* are not lopped but are grazed.

As is expected, the relationship between the density of shrubs and yield of grass is inverse, that is, the high shrub density is associated with less grass yield. Kaul and Ganguly (1963) observed that 14% density of *Zizyphus nummularia* in an *oran* was ideal for obtaining the maximum yield of leaf fodder and grass as compared to 11 or 18% shrub density. The yield and composition of grasses varied under the canopy of different trees. As in the case of agricultural crops, *Prosopis cineraria* was compatible with grasses also, and gave the highest yield of herbage (2.34 tonnes/ha) growing in association with it; the lowest yield was obtained under *Prosopis chilensis* (0.85 tonnes/ha); the herbage yield under *Acacia senegal, Albizia lebbek* and *Tecomella undulata* was 0.78, 1.32 and 1.66 tonnes/ha, respectively (Shankar 1984). It was concluded that *Prosopis cineraria* was the most favourable for the growth of associated grasses and *Prosopis chilensis* was the most unfavourable. Almost similar results have been reported by Saxena (1981), who further reported that *Prosopis cineraria* provided the maximum leaf litter (Table 31).

Table 31: Average plant density, number of species, above ground biomass and litter/sq m under different tree species

Canopy cover	Plant population			No. of species occurring			Above ground biomass (gm)	Canopy density	Litter (gm)
	Perennials (No)	Annuals (No)	Total	Perennials (No)	Annuals (No)	Total			
Acacia senegal	53.2	169.3	222.5	4.7	14.4	19.1	128.9	0.4-0.8	30.3
Tecomella undulata	125.7	225.7	351.4	6.7	16.9	23.6	192.5	0.3-0.6	39.2
Prosopis cineraria	112.5	245.6	358.1	6.7	19.4	26.1	234.8	0.5-0.7	68.0
Prosopis juliflora	52.9	102.7	132.7	3.9	11.7	15.6	117.7	0.4-0.6	23.0
Albizia lebbek	64.3	185.1	249.4	3.8	15.3	19.1	130.8	0.3-0.5	23.4

Source: Saxena, 1981

Zizyphus nummularia starts bearing fruit after the second year of establishment. Yearly cut shrubs on cultivated fields generally do not bear fruit. A five-year-old bush in a grassland may produce 3.5-4.0 kg of fruit; old well-developed shrubs in grassland produce 3-5 kg of fruit in the 150-250 mm rainfall zone (Saxena, 1984). Harvesting of *Zizyphus* leaves starts with the season of leaf fall in winter. It has been confirmed by field studies that winter harvest gives the maximum above-ground biomass leaf and shoot production, productivity per day and other growth parameters of the shrub (Table 32) (Saxena and Sharma, 1981). Extensive information about the mineral content (Ca, K, Mg, Na and P) of the dry leaf fodder (*pala*) in relation to animal requirements (Dutta and Dhir, 1981); distribution of trace elements in *pala* (Sharma and Dhir, 1981) and the nutritive value of pala for ruminants (Bohra and Ghosh, 1980), is available.

Table 32: Growth attributes of *bordi* (*Zizyphus nummularia*) in different seasons

Season	Aboveground biomass (gm)				Plant height (cm)	Basal area (cm)	No. of stumps
	Leaf	Shoot	Production	Productivity per day			
Spring	73.26	348.47	421.73	1.15	141.7	3.55	3.17
Summer	40.03	338.22	378.25	1.04	138.3	4.18	3.35
Monsoon	152.08	248.64	400.72	1.09	146.9	6.02	3.85
Winter	193.09	574.85	767.94	2.10	171.3	7.42	4.07
Mean	114.61	377.54	492.16	1.35	149.6	5.29	3.60
SEm±	34.80	69.87	–	–	5.67	0.61	0.23
F test	Sig	Sig	–	–	H. Sig	H.Sig	Sig
C.D at 5%	100.7	202.15	–	–	16.40	1.76	0.66

Source: Saxena and Sharma, 1981

6.3 OTHER GRASS-TREE MANAGEMENT PRACTICES

Apart from the above grassland and tree management practices, no detailed description of any other such practice is available, even though, livestock graze almost all over the country. In this section the research efforts made to manage the grass and tree systems are discussed with the understanding that they may also focus on various types of prevalent systems. Grassland-tree systems have been improved by various techniques as given here.

6.3.1 Closure to Biotic Interference

It has been observed, that mere closure of an area to biotic interference, restores grass and tree cover indigenous to the area. All the areas in India, due to the prevailing favourable edaphic conditions contain desirable and potentially climax flora which do not get an opportunity to express themselves due to continuous over-grazing and cutting. Once this pressure is taken off, all these flora come up very rapidly, and the grasses achieve a very favourable sub-type, if not the climax type. It is reported that in the Gujarat ravines *Aristida* spp. and *Themeda triandra* were growing in the place of the desirable *Dicanthium-Cenchrus* associate (Tejwani, et al., 1960; 1961). In the ravine lands at Vasad and Kota, as a result of prevention of grazing and other biotic interference, a number of favourable changes took place, such as: (1) *Aristida funiculata* and *Themada triandra* were replaced first by *Apluda mutica* (an annual grass) and then by perennials like *Eremopogon faveolatus, Heteropogon contortus, Dicanthium annulatum* and *Cenchrus* spp.; (2) there was not only a qualitative change in the species but also a quantitive increase in the yield of grasses; and (3) the runoff and soil loss progressively decreased from the area as the natural vegetation improved. The total number of trees and shrubs in the area closed to grazing increased from 313 in 1960-1961, to 349 in 1962-1963, 737 in 1965-1966, and 1,028 in 1973-1974. Thus in a period of 14 years the tree and shrub population increased by 228% and there were 80 trees/ha with a girth at breast height of more than 50 cm (Pradhan, 1977).

In another experiment at Vasad, on 0.1 ha plots, 31 years closure (1959 to 1991) resulted in a dense forest of 372, 222 and 600 trees per hectare on the upland, slopes and bottoms of the ravines, respectively. The canopy was good, ground cover was good, and the ground was covered with leaf litter. Over these years, the barren ravine land passed through the grass phase, the shrub phase, and is now in the tree phase. *Prosopis cineraria* is dominant on the upland, *Azadirachta indica* on the slopes and *Holoptelia integrifolia* in the bottom of the ravines. *Acacia senegal* and *Balanites aegyptiaca* are dominant in all the three locations in the middle storey. Among the lower storey *Capparis decidua, Zizyphus nummularia* and *Triumfetta rotundifolia* are dominant in all the locations (Nambiar, 1992).

Similar results have been reported from the semi-arid ravine areas in Agra, where, in a period of 10 years of closure, edible grass species increased from 45 to 49%, and non-edible species decreased from 43 to 30% (Prajapati, 1974). In the arid zone of Rajasthan, observations taken on a large number of paddocks showed that, after two years of protection and controlled grazing, the forage yield increased by 148, 92 and 116% in the "poor", "fair" and "good" range condition classes, respectively (Bhimaya et al., 1967). For the next four years of observation the grass yield trends were maintained.

Similar observations of increase in the growth of better grass species and grass yield, on the one hand, and better growth of tree species on the other, have been made in the semi-arid Aravalli Hills in the Haryana state.

6.3.2 Plantations

Experiments have been conducted to plant capability class V, VI and VII lands with desirable grass and tree species suited to the prevailing agro-ecological conditions. These experiments were initiated in the late fifties and at that time were called "fodder and fuel plantations" since they were designed to meet these two specific needs of the farmers.

In an experiment conducted in the ravine lands in Vasad where *Cenchrus ciliaris* and *Dicanthium annulatum* were planted separately as well as in association with *Acacia nilotica*, it was observed, that the grasses were compatible with the tree species planted; over a period of seven years of observation, there was no significant difference in the yield of grasses with and without the association of trees. The overall mean grass yield increased from year to year from 3,315 kg/ha in the first year to 6,663 kg/ha in the fourth year, making it a 101% increase in the fourth year of plantation. This increase was partly due to the effect of closure and protection of the area, and partly due to the improvement works of broadcasting grass seeds and planting of grasses. As anticipated, the grass yield decreased as the tree canopy increased (Dayal, 1974). In black clayey soil with a similar annual rainfall (at Kota), there is evidence that in association with *Acacia nilotica* there was not only less yield of grass as compared to alluvial soil (as reported in the above experiment) but the grass yield was also clearly reduced at an earlier stage (four years after planting) as the trees grew. The forage yield decreased from 3,159 kg/ha in 1978 to 786 kg/ha in 1982 (Chinnamani et al., 1982).

In the semi-arid, barren, ravine lands near Agra, it is observed that if the natural grassland of *Heteropogon-Saccharum* type is managed in a pastoralsilviculture practice with *Leucaena leucocephala* and/or *Acacia nilotica* or *Dichrostachys nutan* as a planted tree component, and *Cenchrus ciliaris* or *Cynodon dactylon* as an introduced grass, the yield of harvestable biomass (portion usually consumed by goats which includes tree leaves, small twigs, grass clipped at least 15 cm above the ground) increased from 2.4 tonnes/ha to 6.1-8.0 tonnes/ha in all the treatments with *Leucaena leucocephala*, and 4.4 tonnes/ha with *Acacia nilotica* and *Cenchrus ciliaris* (Sharma and Bhattacharyya, 1988). As against the yield of 12,624 Mcal/ha, and 160 kg crude protein/ha, under natural grassland condition, the *Leucaena* and grass combinations yielded 24,705 to 33,866 Mcal/ha, and 890 to 1,270 kg crude protein/ha; the *Acacia nilotica* and *Cenchrus ciliaris* combination yielded 19,368 Mcal/ha and 580 kg crude protein/ha (Sharma and Bhattacharyya, 1988). These pastures contained five to six trees of *Prosopis cineraria* and eight to nine trees of *Salvadora* spp. per hectare. It may be further noted that the experimental plots were irrigated with saline ground water, which reportedly had limited use for crop production.

In the semi-arid black soil, *Cenchrus ciliaris* gave a much higher yield (3.4 tonnes green grass/ha/cut) than *Dicanthium annulatum* (1.7 tonnes green grass/ha/cut)in association with *Acacia nilotica* trees over a period of six years (Subbayan and Chittaranjan, 1979). During the recent years a number of experiments combining *Leucaena leucocephala* as alley fodder/fuel crop with *Cenchrus ciliaris* grass, in a semi-arid area (Rana et al., 1988), and sub-humid Gangetic plains (Chatterjee et al., 1988) have been started. However since the data reported are for two to three years, it is difficult to evaluate the whole system at present.

It has been reported that in the high rainfall sub-tropical climate at Dehradun, if *Chrysopogon fulvus* grass is planted (at 0.75 × 0.75 m spacing) along with *Dalbergia sissoo* (at 9.14 × 9.14 m spacing) in class V, VI and

VII lands, not only is a high yield of grass (10.55 tonnes/ha/yr) obtained but 97 trees/ha will be available for exploitation after 15 years (Mathur and Joshie, 1972).

In another experiment under similar edaphic conditions at Dehradun, *Chrysopogon fulvus* (a fodder grass) and *Eulaliopsis binata* (an industrial grass used for paper pulp and rope making) have been tried in association with four fodder-fuel tree species, *Albizia lebbek, Grewia optiva, Bauhinia purpurea* and *Leucaena leucocephala.* It was observed that, over a period of four years (1980-1983), when the trees were establishing themselves, an average yield of 3,900-4,400 kg of oven-dry grass/ha/cut was recorded. During the same four-year period, on an average, *Eulaliopsis binata* gave a higher yield (5361 kg of oven-dry grass/ha/cut) than *Chrysopogon fulvus* (3,345 kg oven-dry grass/ha/cut) (Singh and Joshie, 1984).

In the same experiment, which was continued for 10 years (1980-1989), the top feed species, except *Albizia lebbek,* were first lopped at the end of 1984; and *Albizia* was lopped in 1985. All the tree species were lopped as per schedule from 1986 to 1989. Bark fibre was also harvested from *Grewia optiva* from 1986 onwards (Singh et al., 1991). The annual oven-dry grass yield (average of 10 years, 1980-1989) under various tree species was 4,379 kg/ha (under *Albizia lebbek*), 3,849 kg (under *Grewia optiva*), 4,368 kg (under *Bauhinia purpurea*) and 4,316 kg (under *Leucaena leucocephala*). Over a period of 10 years, *Eulaliopsis binata* gave more yield of grass (5,261 kg ovendry grass/ha/yr) than *Chrysopogon fulvus* (3,190 kg oven-dry grass/ha/ yr) (Singh et al., 1991). Among the tree species *Albizia lebbek* had the best and *Leucaena leucocephala* the poorest growth parmeters at the end of 10 years (Table 33). The tree species grew better in association with *Chrysogon fulvus* than with *Eulaliopsis binata* and therefore yielded more fuelwood, lopped fodder and bark fibre (Table 33). Among the tree species, *Albizia lebbek* gave fuelwood 500 kg/ha/yr, and lopped fodder 415 kg/ha/yr (Table 33). *Grewia optiva* gave higher fuelwood yield (585 kg/ha/yr), lower lopped fodder yield (208 kg/ha/yr) as compared to *Albizia lebbek.* However *Grewia optiva* in addition gave 615 kg/ha/ yr of bark fibre. Between these two species *Grewia optiva* is more popular and prevalent in the Himalayan ecosystem. *Leucaena leucocephala* gave the least fuelwood and lopped fodder yield, as was expected in the high rainfall regime in the experimental area. It also appeared that a 50% lopping intensity of trees was generally better than a 75% lopping intensity.

In another experiment under similar agro-ecological conditions, the performance of three grass species grown in association with *Leucaena leucocephala* planted at three densities of population was evaluated for a period of seven years. *Leucaena leucocephala* was planted at 1 × 0.5 m, 2 × 0.5 m, and 4 × 0.5 m spacing (20,000, 10,000 and 5,000 plants/ ha). The grasses (Hybrid napier vr NB5; *Panicum maximum* and *Chrysopogon fulvus*) were planted at 1 × 0.5 m spacing (20,000 plants/ha). It was observed that every year, the maximum yield of *Leucaena leucocphala* was obtained with the maximum density i.e., 20,000 plants per hectare. However, as

Table 33: Growth, yield of fuelwood, lopped fodder and fibre bark of top feed tree species in association with different grass species

Treatment	Growth parameters*		Fuelwood** (air dry) over bark (kg/ha/yr)	Lopped** fodder (oven dry) (kg/ha/yr)	Bark fibre** (kg/ha/yr)
	Height (m)	DBH (cm)			
Main treatments *(tree spp.)*					
T1 *Albizia lebbek*	546	88	500	415	–
T2 *Grewia optiva*	524	71	585	208	615
T3 *Bauhinia purpurea*	423	81	475	315	–
T4 *Leucaena leucocephala*	354	55	298	196	–
Sub-treatments *(lopping)*					
C1 50% intensity	471	79	480	272	708
C2 75% intensity	453	68	451	296	531
Sub-sub-treatments *(grasses)*					
G1 *Chrysopogon fulvus*	486	78	579	366	856
G2 *Eulaliopsis binata*	438	70	356	202	379

Source: Singh et al., 1987b, 1988a, 1990b, 1991
*10 years growth; ** Av. of 4 years (1986 to 1989)

Table 34: Green fodder yield of *Leucaena leucocephala* planted at varying spacings with and without association with grasses (Av. of 7 years 1983–1989)

Treatment	Yield of *Leucaena leucocephala* (kg/ha)	Yield of grass (kg/ha)	Total yield (kg/ha)
Main treatments *(spacing)*			
S1-(1 X 0.5m)=20000 plants	22270	4911	27181
S2-(2 X 0.5m)=10000 plants	15223	6844	22067
S3-(4 X 0.5m)= 5000 plants	11692	8748	20440
Sub-treatments *(grasses)*			
G0-No grass	16342	–	16342
G1-Hybrid Napier(Var NB5)	13031	12882	25913
G2-*Panicum maximum*	13361	6872	20233
G3-*Chrysopogon fulvus*	16076	3299	19375

Source: Singh et al., 1986b, 1987a, 1988a, 1990a

the density of a plantation decreased, the yield of *Leucaena* fodder also decreased (Table 34). The yield of grass was however, the least with the maximum density of *Leucaena*, and it increased as the density of *Leucaena* decreased. Both these observations could be expected, since the yield of grass would be influenced by space, and water available to it in competition with *Leucaena*. As a silvopastoral system, the maximum combined yield of tree fodder and grass was obtained with the maximum density of *Leucaena*. Irrespective of the density of plantation, the sole crop of *Leucaena* gave the maximum fodder yield as compared to its yield in association with grasses. It gave the minimum fodder yield in association with hybrid napier grass. This is explained by the fact that hybrid napier is a very vigorously growing grass. It gave the maximum grass yield (12,882 kg/ha/yr) among the three grasses. A combination of *Leucaena* and hybrid napier gave the maximum fodder yield of 25,913 kg/ha/yr.

In a medium rainfall area with alfisol soils at Jhansi, berseem (*Trifolium alexandrinum*) was grown in association with three tree species, and was applied varying doses of phosphatic fertiliser. The trees were four to five years old and had been planted at 5 × 4 m spacing (Hazra and Tripathi, 1986). As is well established, the application of phosphorus increased the yield of berseem without any association with trees (Table 35). The yield of berseem was very severely reduced when grown in association with *Leucaena leucocephala* and *Acacia tortilis*; the yield decreased in association with *Hardwickia binata* also, but the impact was not very severe. The yield differences in association with the trees were explained by differences in phytosynthetically active radiation (PAR). *Leucaena leucocephala* allowed the least amount of light infiltration which was about 48% of the full light availability in the open; *Acacia tortilis* allowed 65% and *Hardwickia binata* allowed 80% of light availability. As a consequence

Table 35: Yield of berseem forage (kg/ha) as influenced by application of phosphorus and its association with trees (Av. of two years)

Phosphorus fertiliser (kg/ha)	Control (no tree)	Tree species		
		Acacia tortilis	*Leucaena leucocephala*	*Hardwickia binata*
0	2840 (100)	1340 (47)	880 (33)	2440 (86)
21.5	3680 (100)	2210 (60)	1350 (37)	3220 (87)
43.0	4760 (100)	2740 (57)	1830 (37)	4190 (88)
64.5	5100 (100)	3270 (64)	2070 (38)	4490 (88)
CD at 5%	390	290	250	310
PAR*	100	65	48	80

Source: Hazra and Tripathi, 1986

Figures in parentheses refer to per cent yield relative to control

*PAR=Phytosynthetically active radiation, experessed as per cent of the control

Leucaena leucocephala, Acacia tortilis and *Hardwickia binata* gave 36,57 and 78% relative yield, respectively as compared to the crop grown in the open (Hazra and Tripathi, 1986).

In another experiment in the medium rainfall area with alfisol soils, four fodder crops—barley, oats, Chinese cabbage and safflower—were grown under the canopy of *Albizia lebbek, Acacia tortilis* and *Leucaena leucocephala* trees (Hazra, 1985). The age and spacing of the trees are not mentioned; presumably they would have been the same as discussed in the preceding paragraph. It was observed that the yield of forage crops increased as the dose of nitrogen increased (Table 36). Yield of all the forage crops was reduced due to their association with trees; however the beneficial impact of nitrogen was still evident at all the levels of nitrogen.

Table 36: Dry forage yield (kg/ha) of different crops as influenced by N fertilization under different tree species

Nitrogen levels (kg N/ha)	Control (Open)	Tree Species		
		Albizia lebbek	*Acacia tortilis*	*Leucaena leucocephala*
		Barely		
0	2033(100)	1738(85)	1402(69)	1303(64)
40	2754(100)	2385(87)	1770(64)	1549(56)
80	3445(100)	3295(96)	2090(61)	1787(52)
120	3770(100)	3426(91)	2205(58)	2033(54)
CD at 5%	327	327	237	128
		Oat		
0	4279(100)	3893(91)	3131(73)	3025(71)
40	6148(100)	5656(92)	4869(79)	4992(82)
80	7746(100)	7180(93)	6443(83)	6270(81)
120	9016(100)	8459(94)	7213(80)	6885(76)
CD at 5%	875	574	800	659
		Chinese cabbage		
0	1438(100)	1182(82)	1090(76)	1256(87)
20	1885(100)	1576(84)	1373(73)	1543(82)
40	2367(100)	1998(84)	1830(77)	1850(78)
60	2830(100)	2240(79)	2156(76)	1986(70)
CD at 5%	448	241	225	185
		Safflower		
0	2684(100)	1988(74)	2008(75)	1803(67)
20	3361(100)	2561(76)	2766(82)	2500(74)
40	4242(100)	3361(79)	3484(82)	3012(71)
60	4775(100)	3893(82)	3934(82)	3381(71)
CD at 5%	714	446	278	319

Source: Hazra, 1985

Figures in parentheses refer to per cent yield relative to control

The forage yield loss was relatively much less under *Albizia lebbek* than under *Acacia tortilis* and *Leucaena leucocephala*. Among the forages, oats gave the maximum yield of forage, and barley was the most adversely affected. The reduction in the yield of the forages in association with trees was attributed to the reduction in the phytosynthetically-active radiation under the canopy of trees. The average phytosynthetically-active radiation under *Albizia lebbek, Acacia tortilis* and *Leucaena leucocephala* was 80,66 and 58% respectively, as compared to 100 in the open (Hazra, 1985). The lower the phytosynthetically active radiation values under a tree, the greater was the loss of forage yield.

CHAPTER 7

SILVOPASTORAL PRACTICES

The silvopastoral systems concern themselves with the lopping of forest trees and grazing of understorey grasses and bushes in the forests (and also tree plantations). As mentioned earlier, there are 67.1 million ha of forest land in India (MOA, 1992), mostly controlled by the forest departments of the state governments. People have the right to graze their animals in these forest lands and to collect fodder, fuelwood and timber for their needs as per the prescribed rules and regulations. Indian forests are extensively grazed by animals, and fodder-yielding trees are lopped to provide fodder.

7.1 GRAZING PRACTICES

Grazing systems are recognised either on the basis of methods of grazing or on the pattern of migration.

The *Kharak* system of grazing in Uttarkhand in Uttar Pradesh and *Gols* system in Rajasthan are the best examples of methods of grazing. When grass in a *Kharak* or a *Gol* becomes thin or used, the herd is moved, and the herdsmen build other *Kharak* or *Gol* (Randhawa, 1959).

In the case of migratory systems, four patterns are recognised, namely, (1) Total nomadism, (2) Semi-nomadism, (3) Partial nomadism, and (4) Transhumans.

Total nomadism is practised by communities in which the animal owners do not have a permanent place of residence, do not practice regular cultivation, and the whole families move with the herds in search of forage and water. Semi-nomadism is a related system where the animal owners have a permanent place of residence near which supplementary cultivation is practised. However, for the greater part of the year they move with their herds to distant grazing areas. Partial nomadism is practised by farmers who live in a permanent settlement, and keep their herds in its vicinity. The transhumans system is practised by farmers who have a permanent place of residence and whose herds are tended by herdsmen for long periods of time in distant grazing areas (Tyagi and Singh, 1988). Thus the transhumans system is also a totally nomadic system, except that the herds are not owned by the herdsmen. Nearly 30 pastoral communities are recognised in India (Randhawa, 1959). They are concentrated in the northern

(Himalayan) part of India (covering Jammu and Kashmir, Himachal Pradesh and Uttar Pradesh), and the western (arid) part of India (covering Haryana, Rajasthan and Gujarat). They are relatively much less concentrated in south India (Andhra Pradesh, Karnataka and Tamil Nadu). In other states only a few communities are noticed.

7.2 SILVOPASTORAL PRACTICES IN THE HIMALAYA

One of the more outstanding examples of silvopastoral practices is grazing in the forest lands in the Himalaya by farmers and the nomadic communities mentioned above.

In the western and central Himalaya, covering the states of Jammu and Kashmir, Himachal Pradesh and Uttar Pradesh, the forests are divided into those of the tropical, sub-tropical and temperate zones according to the increase in altitude. Above the temperate zone is the alpine zone, which does not have forests.

In the alpine zone rainfall is scanty, although snowfall is good. There are good alpine pastures, and therefore the main vocation is the rearing of sheep and goats.

A good description of the alpine meadows in Uttar Pradesh is provided by Rawat and Rodgers (1988). They suggest a classification into three types of alpine meadows based initially on the aspect which allows integration of presence of species and their ecological adaptation (Table 37). There are subtypes within various aspects depending upon the elevation, as described here.

Sunny slopes (with less moisture due to greater exposure)

a) Danthonia tussock grasslands: This distinctive community is common on steeper slopes between 3,200 and 4,000 m elevation. The tussocks of long needle-shaped leaves reach 50 cm high and are separated by a short herb-rich turf to 10 cm. These areas are locally known as *Fizan* or *met bugiyals*. *Morina longifolia* and *Berberis* spp. are common constituents at lower elevations.

b) *Euphorbia-Gentiana-Anaphalis* turf: This rich and variable community is characteristic of the more gentle slopes from 3,300 to 4,000 m elevation. *Euphorbia stracheyi* is more frequent on the drier exposed sites, known locally as *dud bugiyals*. Local overgrazing causes an increase in herbs such as *Taraxacum* and *Trachydium*.

c) *Kobresia-Astragalus-Potentilla* turf: Another rich and variable community found at slightly higher altitudes from 4,000 to 4,800 m. Several species of *Kobresia* sedge, form a dense and palatable turf known locally as *dhati bugiyal*.

d) *Saussurea-Draba* scattered herb community is typical of areas above 4,500 m elevation, with *Draba* being more common at higher levels and is locally known as *bas bugiyal*.

Table 37: Comparative description of three major *bugiyal* (alpine meadow) types

Features	Sunny	Shady slopes	Marsh meadows
Aspect	South/East	North/West	Flat, water logged, boggy
Meadow subtype	Bas, Dud	Dhanlya Bishan	Naman, Sem
Tree line	3300-3400 m	3500-3600 m	3300-3500 m
Snow line	5400-5600 m	4800-5200 m	–
Edaphic conditions	Dry, gentle to steep slopes	Moist, rugged areas	Along glacial river beds and high, little exposure altitude lake margins, very moist
Scrub fringes	*Berberis – Cotoneaster, Caragana-Artemisia, Juniperus* etc.	*Rhododendron-Lonicera, Sarbus-Lonicera* etc	*Hippophae-Myricaria, Salix* spp. association
Growth Forms	Dwarf, woolly herbs, sedges and grasses	Fleshy herbs with thick rhizomes and tubers, matted shrubs	Scapigerous, marsh-loving herbs, sedges and rushes, more mosses
Grass communities	*Danthonia-Agrostis-*Irisatum (Danthonia type)	*Agrostis-Poa, Phleum-Agropyron* (Agrostis type)	*Daschempsia-Deyuxia, Poa alpina-Phleum* (Phleum type)
Characteristic species	*Saussurea graminifolia, Carex* spp., *Potentilla, Astragalus, Euphorbia stracheyi, Anaphalis, Gentiana*	Phleum *moorcroftianum, Hierochloa, Pleurospermum, Trachydium, Cortia, Aconitum, Polygonum, Geranium, Cassiope*	*Caltha palustris, Triglochin palustre, Eleocharis palustre Phleum alpinum, Scirpus, Parnassia, Suphrasia, Pedicularis, Juncus*
Approx. species	76 genera, 271 spp.	86 genera, 319 spp.	27 genera, 123 spp.
Human/cattle interference	Village at flat bases; more cattle and sheep; overgrazing	Collection of fuelwood, bamboo and medicinal herbs, less grazing.	Grazing during drier months and collection of medicinal herbs

Source: Rawat and Rodgers, 1988

Shady slopes (with less sun and greater moisture)

a) *Rhododendron-Polygonum* tree line shrubberies are not true meadows but are found up to 3,500 m elevation, in the alpine zone. *Rhododendron campanulatum* and *Polygonum polystachyum* dominate over a varied herb layer.

b) *Cassiope-Gaultheria* dwarf shrubberies are common to 4,000 m elevation, sometimes associated with *Rhododendron anthopogon.*

c) *Trachydium-Cortia lindleyi* dwarf turf herb communities are common throughout the subalpine-alpine regions, especially in areas of high grazing pressure. They are known locally as *dhaniya bugiyals.*

d) *Angelica-Aconitum-Pleurospermum angelicoides* herb communities are typical of shady slopes when dwarf shrubs are absent.

e) Grass associations of varied composition are common between 3,000 and 4,200 m elevation. *Poa-Agrostis* are more typical of lower elevation, *Agropyron-Phleum* at higher levels. *Rheum* and *Hierochloa* communities above 4,200 m elevation, are known locally as *teturi bugiyals; Rheum moorcroftianum* more typically on scree slopes and *Hierochloa* on moister sites near melting snow.

Marsh meadows (valley bottoms and boggy areas)

a) *Phleum alpinum-Poa* association is mostly found in over-grazed localities especially near boggy areas between 3,200 and 4,000 m.

b) *Triglochin palustre-Eleocharis palustre-Scirpus* mixed sedge communities are frequent in boggy areas, stream and lake banks, commonly associated with *Juncus* spp.

c) Other herbs—*Caltha palustris, Plantago* spp., *Parnassia* spp., *Primula munroana* etc. are other characteristic herbs growing in such areas. There is a high percentage of moss cover intermixed with *Equisetum* spp. and occasionally *Ophioglossum vulgatum.* The banks of various high altitude lakes, such as Jeolinkong (4,800 m), Anchari *tal* (4,400 m), Suryakund (4,200 m), Zimi *tal* (4,800 m), Hemkund (4,600 m) and Deo *tal* (4,000 m), represent such vegetation, which is more or less stable at various elevations.

In summer, the animals are grazed in these areas, and as the cold weather sets in, the animals move down to the temperate forests, and finally to the subtropical forests. The alpine pastures are reported to be over-grazed in Jammu and Kashmir (Shah, 1988) and elsewhere although no quantitative data are reported. Rawat and Rodgers (1988) report that despite intense grazing pressures, many alpine meadows maintain good vegetative cover, high production and high diversity of species. They are of the opinion that these meadows are grazophilic communities. Degradation of meadows is evident around the permanent camps perhaps more due to trampling than grazing. There is a fairly good sprinkling of medicinal plants such as *Ephedra, Juniper, Artimisia* in Himachal Pradesh. Some medicinal plants (for example, *Aconitum heterophyllum, A. violaceum, Angelica glauca, Dactylorhiza hataegirea,* and *Podophyllum hexanorum*) in the Uttar Pradesh alpine meadows are reported to be over-exploited and hence rare. This is attributed to the complete uprooting of the herbs rather than the harvesting of the leaves (Rawat and Rodgers, 1988).

The temperate zone supports the conifer forests of *deodar* (*Cedrus deodara*), blue pine (*Pinus wallichiana*), and spruce (*Picea smithiana*). Wherever, the pine forest canopy has become thin, grasses dominate the ground floor. Along the streams and depressions broad-leaved species, like maple, cornus, betula, and oak grow. Migratory flocks of sheep, goat and cattle pass through these forests and cause heavy damage to the grass and tree regeneration. This zone is also inhabited, hence the animals of the farming community also contribute to the problem of over-grazing and over-lopping of trees in the forests.

The dominant forest tree species in the subtropical zone are *Shorea robusta, Acacia catechu, Pinus roxburghii, Querqus leucotrichophora* etc. Since the subtropical zone is relatively more densely populated, the pressure of over-grazing, over-lopping, over-cutting and forest encroachment is very high. Consequently, the grass cover is more widely spread in this zone than in the temperate zone. It is reported that farmers in the Renuka microwatershed at an elevation of 700-1,700 m in Himachal Pradesh were short of 1,285 tonnes of dry matter forage (reportedly 24.2% of the total requirement) which had to be obtained from forest tree leaves (Srivastava et al., 1988).

Gaddis, the dominant pastoral community of the central Himalaya, migrate during winter from Palampur in Himachal Pradesh to Ramnagar in Uttar Pradesh via Sujanpur, Hamirpur, Bilaspur, Nalagarh, Pinjore, Bhogriwala, Badshahibagh, Hardwar, Kotdwar and Khalagarh (Tyagi and Singh, 1988). With the onset of summer the route is reversed for going back to the Himalaya.

Sustainability of Forest Grazing

If forest grazing is regulated and practised appropriately, it has many advantages associated with it. For example, it reduces the incidence of herbaceous and short ligneous vegetation, which otherwise becomes fire prone during the dry season; the reduction in the numbers of herbaceous and short ligneous species reduces the competition for moisture and nutrients for the more economic species; the light trampling by cattle provides a mixing of leaf litter, promoting decomposition and mineralisation. In scientific forest management, grazing is recommended both as a technique for proper tree growth and for the regeneration of *Shorea robusta*, sandal wood and teak (Rowntree, 1940; Champion and Griffiths 1948). However, at present over-grazing is more of a rule than otherwise. Over-grazing causes much more severe damage. For example it leads to the elimination of valuable tree seedlings, creates deformities in trees due to repeated browsing and lopping, compaction of soil, acceleration of runoff and erosion, etc. Not much work has been done on the carrying capacity in forests. Some estimates of carrying capacity were reported in as far back as 1946 for forests in the erstwhile Bombay Presidency, which were based on rainfall. These were (Kumar, 1946)

Rainfall (mm)	*Animals per ha*
>2540	0.2 - 0.3
1270 - 2540	0.4 - 0.6
890 - 1270	0.8 - 1.0
653 - 890	1.2 - 1.6
<510	1.6 - 2.0

7.3 FOREST TREE LEAF FODDERS

Forest tree leaf fodders have always been used by nomadic graziers and farmers. In view of this, whenever forest trees have been described, their utility or otherwise to provide leaf fodder is also noted (Brandis, 1906; Talbot, 1909, 1912; Troup, 1921; Gorrie, 1937; Drummond, 1944; Chaturvedi, 1948). As many as 33 tree speceis are lopped for tree fodder in the central Himalaya (Singh et al., 1988c) at an elevation of 300-3,100 m (Table 38). Tree leaf fodder is available between October and July. Tree leaf fodder is most valuable in the dry season. There is no need to lop the trees during the rainy season—July to September.

Important forage grasses occurring at different elevations and associated with different forest types, subalpine and alpine communities are reported in Table 39. The dominant grasses are marked with an asterisk (Singh et al., 1988c).

Gupta (1981) has recognised 59 species of tree leaf fodders used by farmers in the northeastern hills of India. *Artocarpus, Bauhinia, Careya, Ficus, Grewia* and *Vitex* species are reported to be the most important for providing leaf fodder to the domestic livestock during the lean season of the year in the northeastern hills (Varma et al., 1982, a, b). An analysis of 580 bundles of fodder purchased from the farmers in the month of August indicated that they contained on an average 55% grass species and the balance comprised fodder herbs, shrubs, trees and ferns. As many as 21 grass species were being cut. The dominant grass species being cut were: *Arundella bengalensis, Bambusa* spp., *Thysanolaena maxima, Imperata cylindrica, Setaria pallidefusca, S. palmifolia, Eragrostis cilianensis,* and *Erianthus longisetosus.* Among the non-grasses as many as 27 genera belonging to 13 families were being utilised (Varma et al., 1983).

No attention whatsoever has been paid by the scientists to the lopping practices of the trees in the Himalaya. Gorrie (1937) mentioned that all the six oaks common in the western Himalaya—*Quercus glauca* in the lower hills, *Q. leucotrichophora, Q. dilatata* and *Q. lanuginosa* in the middle elevations, and *Q. semecarpifolia* in the higher elevations were traditionally lopped on a regular rotation of two to three years cycle. Even then, that is over half century ago, he recorded that the growing population, the reservation of large tracts as government forests, and ill-defined but liberal

Table 38: Common fodder tree and shrub species of the central Himalaya, their elevational distribution and months of use

Species	Elevational range (m)	Use-period (months)
Acer pictum	2000-3000	Jan-Mar
Bauhinia retusa	300-1200	Dec-Feb
Bauhinia variegata	300-1800	Dec-Jan
Boehmeria rugulosa	450-1650	Dec-Feb
Carpinus viminea	1600-2600	Nov-Feb
Castanopsis tribuloides	1000-2200	Jan-May
Celtis australis	1650-2700	May-June
Cornus macrophylla	1500-2400	Oct-Feb
Crataegus crenulata	700-2300	Feb-June
Daphniphyllum himalayense	1700-2600	Oct-Feb
Debregeasia hypolenca	1000-2300	Feb-June
Desmodium spp.	1500-2200	Nov-Jan
Ficus clavata	500-1300	Feb-May
Ficus faveolata	700-2500	Dec-Apr
Ficus infectoria	500-2200	Dec-Mar
Ficus micrantha	450-1650	Dec-Jan
Ficus nemoralis	1200-1950	May-June
Ficus palmata	300-1800	June-July
Grewia elastica	300-1700	Oct-Feb
Grewia optiva	300-1200	Dec-Feb
Melia azedarach	300-1800	May-July
Morus serrata	1200-2600	Nov-Dec
Osmanthus fragrans	1200-2200	Dec-Feb
Phoebe lanceolata	500-1500	Oct-Dec
Quercus glauca	1000-2200	Dec-Mar
Quercus floribunda	2000-3000	Nov-Feb
Quercus lanuginosa	1400-1800	Nov-Feb
Quercus leucotrichophora	1500-2200	Nov-May
Quercus semecarpifolia	2300-3100	May-Oct
Strameaesia glaucescens	900-1600	Nov-Jan
Urtica dioica	1000-2200	May-Oct
Woodfordia floribunda	600-2200	Jan-Oct
Zizypus mauritiana	300-1000	May-Aug

Source: Singh, et al., 1988c

rights and privileges, had led to increasing disregard for the continued productivity of these trees. If the situation was breaking down then, it is but certain that the trees are currently over-lopped, the land over-grazed and the system biophysically disintegrated.

The leaf fodder tree species of the semi-arid and arid regions have been described in sections 2.11 and 6.2.

Table 39: Major grass species at different elevations and related forest types in the Himalaya

Forest type and elevation	Dominant grass
Chir pine (*Pinus roxburghii*) and mixed 350-2000 m	*Heteropogon contortus**, *Chrysopogon serrulatus**, *C. fulvus**, *Cymbopogon distance**, *C. stracheyii*, *Bothriochloa intermedia**, *Alpuda mutica*, *Dicanthium annulatum**, *Cynodon dactylon*, *Eulaliopsis binata*, *Themeda anathera*, *Poa annua**, *Koeleria cristata*, *Muhlenbergia duthiana*, *Festuca kashmeriana*, *Panicum trypheron*, *Paspalidum flavidum*, *Pennisetum orientale* and *Setaria glauca*
High altitude 2000-3000 m	*Agropyron longearistatum**, *Agrostis pilosula*, *Andropogon tristis**, *Anthraxon serrulatus**, *Brychypodium sylvaticum*, *Calamagrostis pilosula*, *Dactylis glomerata**, *Poa pratensis*, *Stipa sibirica*, *Deyeuxia scabrescens*, *Phleum alpinum*, *Poa alpine** and *Sporobolus* spp.
Subalpine and alpine Above 3000 m	*Danthonia cachemyriana**, *Koeleria cristata*, *Calamagrostis emodensis**, *Festuca lucida*, *Brachypodium sylvaticm**, *Trisetum spicatum*, *Andropogon tristis*, *Phleum alpinum**, *Agropyron semicostatum*, *Agrostis munroana*, *Deschampsia caespitosa*, *Helictotrichon virescens* and *Deyeuxia pulchella*

Source: S.P. Singh et al., 1988c

*: Dominant grass species

7.4 NUTRITIVE VALUE OF TREE LEAF FODDERS

Since tree leaf fodders are extensively fed to livestock, their nutritive value has received considerable attention of animal nutritionists. The earliest treatise on animal nutrition in India contained information on the chemical composition and nutritive value of leaves of some tree leaf fodders (Sen, 1953). An extensive information on the chemical composition of 75 tree leaf fodders was published by Sen et al. (1978). An analysis of additional 21 species is reported by Shankar et al. (1988). The information given concerns crude protein, crude fibre, nitrogen-free extract, ether extract, digestibility coefficients and nutritive values (Appendix 3).

Crude protein includes true proteins, and non-protein nitrogenous compounds such as amides. Ether extract includes fats. Crude fibre includes cell walls and woody fibre and is the least digestible part of the fodder. Nitrogen-free extract includes carbohydrates—starch and sugar.

The value of fodder, or for that matter any foodstuff, depends on the proportion of it, which is digested. The ratio of the digestible portion of the total amount fed, multiplied by 100 represents the digestibility coefficients of a particular nutrient. Most of the digestible coefficients reported in Appendix 3 were determined on adult non-producing stock (Sen et al., 1978). These coefficients are liable to be lower in producing animals due to their large feed consumption. The processing of roughages and concentrates influences the digestibility of nutrients. For example, while the grinding, chopping and chaffing of roughages increase the feed intake, at the same time they increase the rate of passage of such processed food.

From the data of the chemical analysis and the digestibility coefficients (Appendix 3), the nutritive values in terms of digestible crude protein (DCP), total digestible nutrients (TDN), starch equivalent (SE), digestible energy (DE) and metabolisable energy (ME) have been calculated. Digestible crude protein fraction represents for all practical purposes, the available protein in a food. Starch equivalent is the potential energy-producing capacity of a food, expressed as kilogrammes of digestible starch per 100 kg of the material. The basis of calculation of SE, is rather complex. For example, 1 kg SE = 1 kg digestible carbohydrate = 1.07 kg digestible protein = 0.53 digestible fat. In the case of roughages, a considerable amount of energy is spent in chewing, ruminating and other associated work of digestion. For roughages, therefore, a correction is made for the fibre content by deducting graduated values from the total SE values for carbohydrates, protein and fats. In case the crude fibre percentage is less than 4, then 0.29 is deducted for each percent of actual crude fibre; this value increases gradually and progressively to 0.58 when the crude fibre is 16% or less.

Nutritive ratio (NR) is the ratio of digestible protein to digestible non-nitrogenous nutrients. The amount of non-nitrogenous nutrients is reckoned as the sum of digestible carbohydrates, and digestible ether extract multiplied by 2.25. The value of the nutritive ratio gives the idea of the proportion of digestible protein in relation to other nutrients. This ratio is narrow in case a fodder is rich in digestible proteins, and wide in case it is rich in carbohydrates and/or fats.

The gross energy (GE) of a foodstuff is the energy given out in the form of heat when it is completely burnt to its ultimate oxidation products, carbon dioxide and water. The gross energy (GE) less the energy lost in faeces is known as the digestible energy (DE). A portion of the digestible energy (DE) is lost due to the bacterial fermentation of crude fibre as methane and carbon dioxide and is excreted by animals either through the rectum, or by belching and expiration; a portion of energy is lost in the urine. When the losses of energy through urine and fermentation are subtracted from the digestible energy (DE), the net energy available to the animals is known as metabolisable energy (ME). The ME values reported in Appendix 3 have been calculated by multiplying TDN values with a factor that 1 gm of TDN is equivalent to 3.6 Kcal of ME in the case of ruminants (Sen et al., 1978).

The mineral content—(total ash, Ca and P_2O_5) of the tree leaf fodders is reported in Appendix 4. Minerals are important in the nutrition of animals particularly those which are either in the active stage of growth, producing milk, working, or pregnant.

The proximate analysis of the feeding stuffs, which has been reported above, partitioned carbohydrates into crude fibre (CF) and nitrogen-free extract (NFE). This partitioning was based on the assumption that crude fibre was less digestible and nitrogen-free extract was more digestible. There has been a criticism of this partitioning since neither the crude fibre nor the nitrogen-free extract represents any particular chemical constituent or a group of constituents. They vary in composition with different plant species, stages of maturity of a plant, and the conditions under which they are determined. The nitrogen-free extract contains lignin also (Sen et al., 1978).

The knowledge of limitations associated with the determination of crude fibre has stimulated interest in other systems of analysis of feeding stuffs. Van Soest (1967) suggested a detergent system of analysis which partitions the nutrients in crude protein (CP), neutral detergent fibre (NDF), cell content (CC), acid detergent fibre (ADF), hemicellulose (HC), lignin, cellulose and silica.

The cell content (CC) includes sugars, starch, organic acids, lipids, pectins, non-protein nitrogen, soluble pectin which are almost completely digestible in ruminants. The cell wall contains cellulose and hemi-cellulose (which are partly digestible), and lignin, cutin, silica and heat-damaged protein (which are indigestible) (Pradhan and Gupta, 1988). The chemical composition of some tree leaf fodders as per the detergent system of analysis is reported in Appendix 5.

7.5 PALATABILITY OF TREE LEAF FODDERS

Apart from the nutritive value of a fodder, its palatability is also of importance. The relative palatability of various tree fodders, or of the same fodder at different seasons or at different stages of growth has also received attention. Clousten and Plymen (1920) observed that leaves of small bamboo were much relished by cattle. Branford (1922) reported that cattle relished leaves of *Dalbergia sissoo*, and no adverse effects were noticed on feeding them; villagers however, believe that milk of cows and buffaloes dries up when they are fed with *Dalbergia sissoo* leaves (Sen, 1953). Momin and Ray (1943) reported that the relative palatability of nine tree leaf fodders in the descending order was *Zizyphus* spp. (*beri*), *Dendrocalamus strictus* (*bans*), *Ficus religiosa* (*pipal*), *Ficus* spp. (*paker*), *Dalbergia sissoo, Azadirachta indica, Syzygium cuminii, Tamarindus indica,* and *Mangifera indica. Zizyphus* and *Ficus religiosa* were found to be relished by animals throughout the year, but others were eaten in quantity only at their more succulent stage. The palatability of *Celtis australis* (Sharma and Gill, 1969) and of *Dalbergia sissoo* decreases as the leaves mature (Momin and Ray, 1943). In some cases young leaves contain toxic substances, for example, *Robinia pseudacacia* (Pal et al., 1979).

Palatability varies with animals also. For example goats relish the leaves of *Ficus bengalensis* more than bullocks, and the former consume two-and-a-half times the weight of leaves as compared to the latter (Mia et al., 1960). Goats not only consume more leaves but also digest and utilise them more efficiently than bullocks. For example, with goats the TDN, SE values (kg 100 dry matter) and nutritive ratio were 51.1, 35.1 and 11.5, respectively, while the same values for bullocks were 44.5, 28.6 and 21.5 (Mia et al., 1960).

7.6 IMPROVEMENT OF SILVOPASTORAL PRACTICES

Due to an increase in the human and livestock population in India, and the consequent increased demand for fuelwood, timber, grass and tree leaf fodder, trees, whatever and wherever available, are severely degraded. F example, the extent of degradation of the forest in a subtropical to tempe ...te watershed of 370 ha can be gauged by the fact that out of 133 ha of forest land in it, 26% forest land had no canopy, 12% had a thin canopy, only 62% forest had a moderately dense canopy, and there was no forest with a dense canopy (CSWCRTI, 1981). This is not an isolated example but is generally true of the western and central Himalaya.

Experiments have been conducted to grow grasses along with densely planted *Eucalyptus* spp. Basically, the *Eucalyptus* does not encourage the growth of grasses in low rainfall areas; but in areas with more than 1,000 mm of rainfall, preferably in the southwest monsoon, as well as in winter, some grasses grow under it. In this case the goal has been to have a fuelwood tree plantation and to obtain some grass yield as a bonus. The situation is similar to intercropping with coconut or with horticultural trees. In one experiment, tall plants (2 m and above) of *Eucalyptus hybrid* were planted at a spacing of 2 × 2 m. To promote moisture conservation, tie ridges of 15 cm height and 2 × 2 m spacing were also made. *Eulaliopsis binata* grass was planted with *Eucalyptus hybrid.* Within a period of eight years, the trees had attained an average height and girth of 14.7 m and 38.2 cm respectively. During a period of eight years 45.2 tonnes of air-dry grass/ha and during a period of eight-and-a-half years, 179 tonnes of air-dry tree biomass/ha were obtained (this comprised 148 tonnes of poles, 19 tonnes fuelwood and 12 tonnes leaf). When the grass was grown alone it yielded 114.4 tonnes air-dry grass during a period of eight years. After the *Eucalyptus* was harvested, the yield of grass spurted and 9.8 tonnes of air-dry grass/ha/yr were obtained in *Eucalyptus* plot as compared to 13.6 tonnes in the exclusive grass plot during the first year after coppicing (Mishra and Sud, 1983). In the second year after coppicing the *Eucalyptus* plants grew very vigorously and coppice shoots attained a height of 7 m and girth at breast height of 14.9 cm. The grass yield in *Eucalyptus* plots had come down to 4.25 tonnes/ha, reflecting the interaction with the shade of coppice shoots (Mishra and Sud, 1984). At the end of four years after harvesting, the *Eucalyptus* coppice shoots had attained an average height of 9.23 m and a girth of 23.4 cm (Sud and Mittal, 1986). Since *Eulaliopsis* grass is used for paper making (October cut) and as fodder (June cut), it fetches high economic returns. Grewal et al. (1992) have reported a total return of Rs. 4,672/ha from the sale of grass from a four-year *Eucalyptus* + grass plantation, as compared to Rs. 1,679 from agricultural crops in 1989.

In another experiment at Dehradun, *Chrysorpogon fulvus* fodder grass was grown in association with *Eucalyptus hybrid.* It was observed that over a period of four years air-dry grass yield of 3.9, 3.4 and 3.9 tonnes/ha/year was obtained when *Eucalyptus* was grown at a spacing of 3.5 × 3.5 m, 5.0 × 5.0 m and 7.0 × 7.0 m, respectively (Puri and Joshie, 1979).

Attempts have been made to evolve silvopastoral practices in the tropical humid climate (mean annual rainfall 2,518 mm, and mean daily minimum and maximum temperature of 23.9 and 31.3° C respectively) by cultivating fast-growing tree species (*Leucaena leucocephala, Casuarina equisetifolia, Acacia auriculiformis* and *Ailanthus malabarica*) at 4 × 4 m spacing, and four different grasses (*Pennisetum perpureum, Panicum maximum, Brachiaria ruziziensis and Euchalena mexicana*). It was observed that the growth and yield of grasses was adversely influenced by the tree component only when the tree canopy was formed (after three to four years). All the grass species gave a high forage yield in association with *Casuarina equisetifolia* and *Ailanthus malabarica* even after the canopy formation. All the grass species gave less forage yield under *Leucaena leucocephala* and *Acacia auriculiformis. Casuarina* with *Pennisetum/Panicum* was observed to be an optimal tree and forage combination (Mathew et al., 1992).

CHAPTER 8

AGROSILVOPASTORAL PRACTICES

Since, Indian farmers, makes use of animals for both draught and milk, most of the agrosilviculture and silvoagriculture practices would be classified as agrosilvopastoral practices. The general structure of these practices is depicted in figure 14. Some of the practices, which do not have animals as a component, are commercial crops of tea, coffee, and cardamom grown under the shade of trees. The animal component is variable, for example, the practices of trees in agricultural fields and on farm boundaries have relatively more dependence on animals, than the practice of growing agricultural crops in combination with commercial trees. Even in shifting cultivation, there is an animal component of pigs and poultry. However, since the animal component has not been studied in all these practices, there are not many facts and data available to describe these as agrosilvopastoral practices.

Even some of the silvopastoral practices would qualify to be agrosilvopastoral practices when they interface with agriculture. For example, when nomadic herds of animals, which depend entirely on a silvopastoral practice, are permitted to pass the night on agricultural lands with a clear understanding that the dung and urine dropped on the land would benefit agricultural crops, and the animals would graze any agricultural waste on the lands, the practice in a strict sense becomes agrosilvopastral. Recently one such practice has been described from the Saurashtra region of Gujarat (Cincotta and Pangare, 1993). This practice results in nutrient and energy trade between the farmer and the pastoralist, and in the process, supports agriculture on the one hand, and compensates for the loss of common grazing land and fertiliser needs.

The practice of home gardens in Kerala and Tamil Nadu combines animals, birds and fish. It is a truly agrosilvopastoral practice.

8.1 HOME GARDENS

The home garden system is practised very extensively in the tropical, high rainfall areas in India, Sri Lanka, Thailand, Malaysia, Indonesia, Papua New Guinea, and in South and Southeast Asia. As shifting cultivation is a response of tribal people, (who do not practise sedentary agriculture), to specific

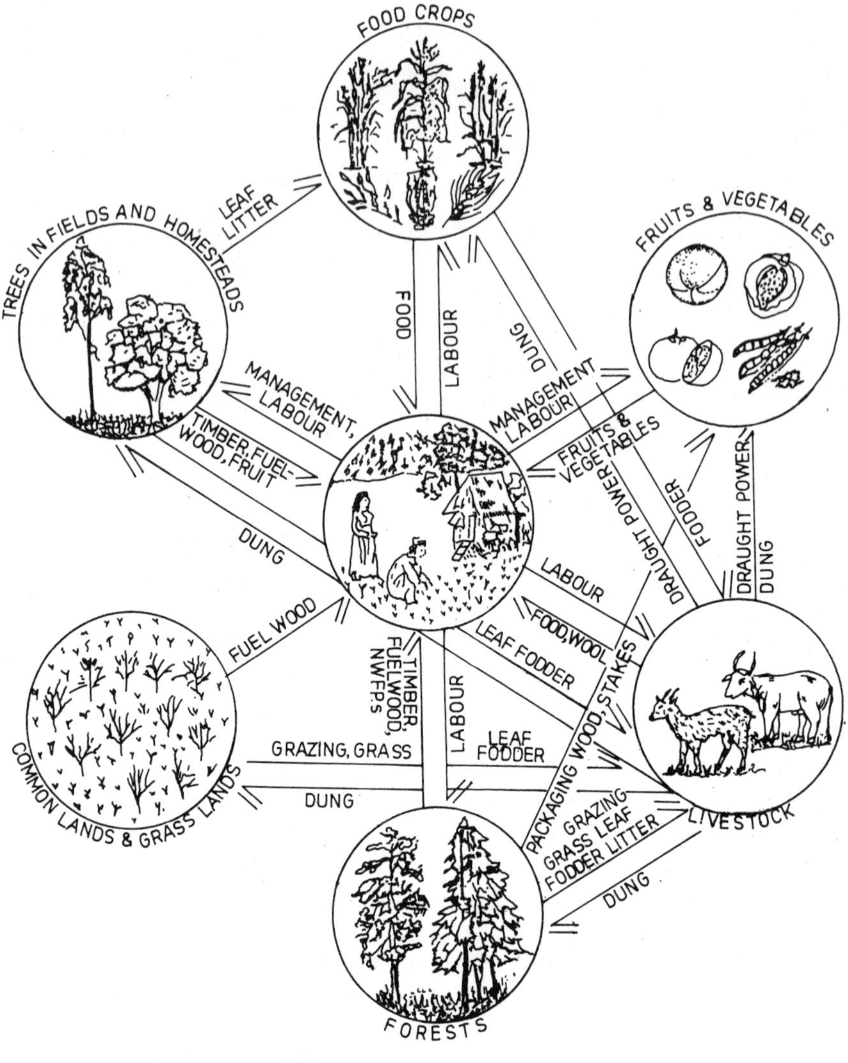

Fig. 14: Structure of agrosilvopastoral systems (Source: Singh, 1991)

high rainfall and other edaphic conditions in the tropical areas, home gardens also are a response to the similar climatic and edaphic conditions by the people who practise sedentary agriculture. Wiersum (1982) has described different types of tree gardens. He has defined a home garden as a "land-use form on private lands surrounding individual houses with a definite fence, in which several tree species are cultivated together with annual and perennial crops, often with the inclusion of small livestock". These tree gardens in Sri Lanka are called "Kandy Gardens" or "Kandy Home Gardens" since this practice of tree cropping has found its maximum potential expression under the present level of management in the district of Kandy.

Distribution

This practice is more popular in Kerala and Tamil Nadu, which have a tropical humid climate, and where coconut is the main crop. It is also found in Orissa. The practice of home gardens is built mostly around coconut. The edaphic requirements for home gardens are similar to those for coconut and arecanut; these have already been described earlier. Since Kerala accounts for the maximum area and production of coconut in India, the home garden practice also has its maximum potential and expression in Kerala. It is only very recently that any description of home gardens in India has become available when Nair and Sreedharan (1986), have described the home gardens of Kerala.

The land area of Kerala can be broadly divided into three physiographic regions—highland, midland and lowland—running almost parallel to each other, along their length from north to south. The mountainous land of the western ghats on the east of the state is the "highland"; it is under forest interspersed with small streams; this region accounts for 48% of the area and 15% of the population of Kerala. The coastal belt on the west, running parallel to the Arabian Sea, with level topography, constitutes the "lowland" (10% area and 26% population). Between the lowland and the highland, lies the "midland" region (42% area and 59% population) interspersed with numerous rivers, small hills and valleys.

Edaphic Conditions

Kerala state as a whole has a humid tropical climate. The rainfall is bimodal, received from the southwest and northeast monsoons; the former accounting for the major share of the total annual rainfall. The mean annual rainfall is 2,960 mm, with a mean annual number of 126 rainy days. The mean annual temperature of the state is 27° C.

The soils of Kerala are broadly divided into four major classes: oxisols (50%), inceptisols (25%), entisols (20%) and alfisols (5%). The state is divided into the following eight major agroclimatic zones on the basis of topography, soils and seawater intrusion: southern, central, northern, high altitude, Onnatukara, Kuttanad and Kole, Pokkali, and low rainfall zones.

The Practice

The cropping systems of Kerala are recognised in four major groups: (1) coconut based, (2) rice based, (3) cassava based, and (4) homestead based. A homestead is an operational farm unit in which a number of crops (including tree crops) are grown along with livestock, poultry, and/or fish production, mainly to satisfy the farmers' basic needs. It is an age old practice and is prevalent throughout Kerala.

Crops

An inventory (Appendix 6) of the major agricultural crops, horticultural trees and forest trees grown in home gardens in Kerala is reported by Nair and Sreedharan (1986). They have reported the presence of eight food species, four pulse species, 14 fruit species, three oil and fat species, two beverage species, eight spice and condiment species, eight vegetable species, 45 fodder grass species, two oil-yielding grass species, and 13 tree species of varying uses—a total of 66 species. Sathees Babu et al. (1992) have reported a total of 60 species in a home garden from South Kerala with a rainfall of 3,000 mm and oxisol soils. Apart from the various uses of the species described above, Sathees Babu et al. (1992) recognised 18 species of ornamental plants. They further reported the number of each of the 60 species in a 0.40 ha home garden. The home garden was coconut-based with 55 trees; among the crops, ginger was the most dominant with 500 plants, occupying 20% of the ground storey. Altogether, there were 848 plants per 0.4 ha. Although home gardens appear to be random mixtures of trees, shrubs and herbs, to a general observer, one can be certain that the farmers know what it is all about.

Wide variation in the intensity of cropping is noticeable between home gardens situated in the same agro-climatic zone. This can generally be attributed to the differences in the socio-economic conditions of the farmers, and their response to externally-determined changes, particularly prices of inputs and products, dependence on land tenurial conditions, etc. It is observed (Table 40) that as the pressure of population increased, and the size of holdings decreased, the intensity of cultivation increased (Nair and Krishnankutty, 1984). In such a situation, it is commonly noticed that as the intensity of tree cropping is increased, miscellaneous tree species, promising no immediate benefit, are replaced by multiple-use tree species. Although this reduces the diversity of species, the intensity of cultivation/use of the selected species increases considerably.

Table 40: Size of holdings and tree intensity in the home gardens of Kerala state

Size of holding (ha)	Mean tree cropping intensity*	Mean no of trees/ha
Very small (0.02–0.19)	50.52	620.8
Small (0.2–0.80)	37.79	276.8
Medium (0.81–2.00)	26.54	211.7
Large (more than 2)	23.24	121.1

Source: Nair and Krishnankutty, 1984

*: Intensity expressed as the percentage of total area occupied by the tree canopies.

Coconut is the most dominant tree species. The other important commercial perennial crops are arecanut, black pepper, cacao and cashew. In this practice, coffee is intercropped with perennial crops like black pepper, jack fruit and others.

Among the tropical tubers, cassava and yam are the most common. Other tuber crops like taro, elephant foot yam, dioscorea, lesser yam, arrow root, sweet potato and Chinese potato are also grown. Banana, pineapple, guava, papaya, annona, sapota, mango and jack fruit are common fruit crops in most home gardens.

A number of pulses (beans), like pigeon pea, cowpea, black gram, and horse gram, and oil seeds,such as groundnut, sesamum and castor, are common in home gardens.

A number of multipurpose trees having productive and protective functions are integrated in the practice and are generally planted on plot boundaries. Teak (*Tectona grandis*) is common in the midlands; a few wild jack fruit trees (*Artocarpus hirsuta*), and *Casuarina equisetifolia* as a wind break are common in the coastal tract of the northern zone; silver oak (*Grevillea rubusta*) is common as a shade tree for coffee in the high range region; bamboo (*Bambusa arundinaea*) is commonly grown along the plot boundaries in the central and northern zones; portia tree (*Thespesia populnea*), numbering five to 10 per home garden, is popular in the southern and central zones. Soft wood tree species, like *pala* (*Alstonia scholaris*), *matty* (*Ailanthus triphysa*), and *kapok* (*Ceiba pentandra*) are planted along the homestead boundaries. Of late, *Albizia falcataria* is being planted in the southern region.

Mangroves form an integral part of the homesteads of the backwater areas in the lowlands of the Pokkali zone, in the Kuttanad zone and the coastal tract of the northern zone. The mangroves protect the homesteads from mud and sand washed down by the rivers and also from high tides. Often they constitute boundaries between homesteads. Mangrove species like *Acanthus ilicifolius, Avicennia officinalis, Carbera odollam, Rhizophora conjugata* and *R. mucronata* are commonly seen.

Animals

Most homestead gardens support a variety of animals—cows, buffaloes, bullocks, goats, sheep, and birds like chicken and ducks. In certain places pigs are also reared. A homestead usually owns one or two cows or buffaloes and 10 to 20 chicken. In the homesteads of backwater areas and coastal tracts, ducks are common. Piggeries are found in the homesteads of central Kerala. Each family invariably owns at least one pair of bullocks, but now the number is dwindling because of the high cost of maintenance.

Fodder and legumes are widely grown to meet the daily fodder requirements of cattle; the important fodder grasses are Guatemala grass, napier grass and Guinea grass (Appendix 6). The waste materials from crops and homes are used as fodder/feed for animals/birds and livestock wastes are used as manure for crops.

The important fish grown are, catla (*Catla catla*), *rohu* (*Labeo rohita*), *mrigal* (*Cirrhina mrigala*), and pearl spot (*Etroplus suratensis*); important prawns are Indian prawn (*Penaeus indicus*) and tiger prawn (*Metapenaeus indicus*).

Bee keeping may also be practised by the more enterprising farmers.

Arrangement of Components

All the crops are not grown in any specific pattern or design but are apparently grown in a scattered way, and are intricately mixed. However, it is difficult to accept that a system which has evolved over a long period of time, on which a farm-family subsists, and which is sustainable could be casual about the location, spacing and site conditions of cash and perennial crops. It may be assumed that those who practise it, know in a practical way, what, where and when to plant, harvest and remove. It will also be difficult to devise an acceptable index of cropping intensity for a situation where crops grow at several tiers above the land, for example, food crops, vegetables and pulses at the ground level; coffee, cacao and many shrubs at a lower height, and finally all of them under the canopy of coconut, arecanut and other tall trees in a recognisable, but not in any specific order. Wherever coconut is the dominant component, the intercrops vary according to the age and canopy of coconut palms as described under intercropping with coconut practice, earlier in this book.

Interactions between Components

No information is available on this aspect of home gardens. Obviously, there would be many interactions among and between the plants and their environment. What distinguishes this practice, from any other practice, is the fact that the plants must explore and utilise the soil and the above-ground space in all the three directions. It is also possible that in a mix of many crop, shrub and tree species, the risk of crop pests and diseases associated with monocropping systems are minimised. However, all this is in the realm of rationalisation as there are no data available to prove or disprove any of these conjectures.

However, there is a very great socio-economic interrelationship and dependence of various components among themselves. This is perhaps best expressed by figure 15.

Resource Input and Utilisation

On an average, each household consists of six to eight persons who provide the workforce for the home garden. Most of the crops in the home gardens are labour-intensive except coconut. As compared to monocropping, the home gardens involve high inputs; for example a home garden with an intensive crop mix and animals requires about 1,000 mandays as compared to 150 days for coconut monocropping and 400 days for paddy monocropping. During certain seasons, such as sowing and/or harvesting, an acute shortage of labour is a common experience.

Since the primary objective of a home garden is to produce a range of commodities primarily for consumption by the farm family, all the operations leading to the sustenance of a home garden are performed within and around the family. No analysis has been done in India to apportion the broad spectrum of activities relating to home gardening and management to various family members. However, some information on this aspect is available about the home gardens (Kandyan home gardens) in Sri Lanka

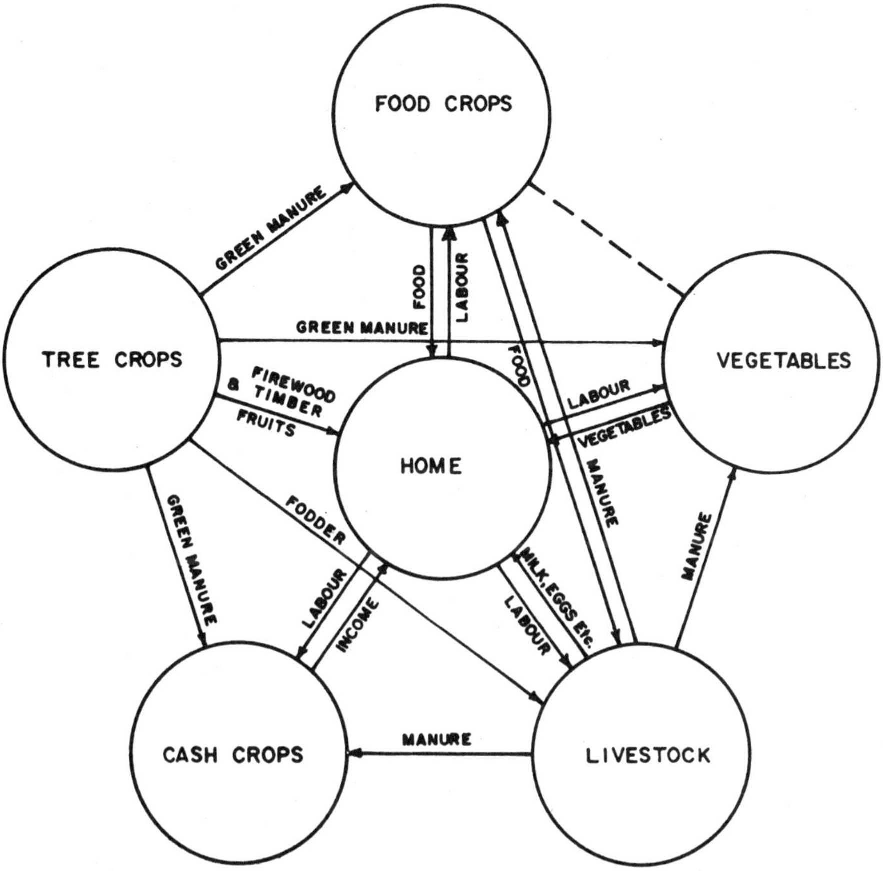

Fig. 15: Interrelation of various components of home gardens among themselves and needs of homes (Source: Nair and Sreedharan, 1986)

(Wickramasinghe, 1991) wherein various activities have been divided into six categories: (1) establishment of the predominants (trees), (2) maintenance of trees and crops, (3) enrichment, (4) harvesting, (5) processing and utilising, and (6) marketing. The role played by the men, women and children in the various activities is reported in table 41. It is obvious that in the case of trees the primary role of women is in the areas of planting, tending, enriching, processing, and utilising; while women share the responsibility with the men in establishment. In the case of tubers, herbs and vegetables it is the women who are primarily responsible in all aspects.

Cash expenses are the maximum, and cash income is the minimum in July-August; this period coincides with the southwest monsoon when planting and cultural operations of most of the crops take place; added to this, the schools open at this time of the year when the farmers have to spend a lot of money for at least two school-going children per family. To meet the immediate cash requirements the farmers try to sell all

Table 41: Home garden development activities and sources of labour

Activity	Plant form		
	Trees	Tubers	Herbs and vegetables
Establishment	Men, Women	Women, Men, Children	Women, Men, Children
Planting/Germination	Women, Children	Women, Men	Women
Sources of planting/ sowing materials	Natural, Neighbour, Relative, Unknown	Women, Men	Women
Source identification	Men, Women	Women, Men	Women, Men
Management decisions	Women, Men	Women	Women
Management activities	Men, Women	Women	Women
Enriching the system	Women, Men	Women, Men	Women
Decisions	Women, Men	Women, Men	Women
Harvesting decisions	Women, Men	Women	Women
Harvesting	Men, Women	Women	Women
Processing	Women	Women	Women
Utilising	Women	Women	Women
Marketing decision	Women	Women	Women
Marketing	Men, Women	Men, Women	Men, Women

Source: Wickramasinghe, 1991

disposable farm animals and also borrow money either from village money lenders (at excessive interest rates), or cooperative banks, or agricultural credit societies. Apart from cash, inputs like fertilisers, implements, tractors, pumpsets can also be obtained through these societies.

Seeds are usually stored by the farmers for use in the ensuing season. Seeds/seedlings are also obtained from the authorised government agencies and private suppliers. Chemical fertilisers are being increasingly used, but plant protection chemicals are not widely used.

The major commodities marketed are arecanut, cacao, coconut, coffee, cashew, ginger, pepper, rubber, turmeric and vegetables.

Production

The production values, per unit adopted, are reported in Appendix 6. Nair and Sreedharan (1986) have not commented on the relative yields, and their interrelationship in the home garden system. However, depending on the mix of crops and animals, one could expect a steady harvest/income throughout the year except for the months of July and August. Since, no case studies are reported, it is difficult to evaluate the production pattern. To evaluate the variation in individual farms a "Time Concentration Index" has been suggested. The values of this "Time Concentration Index" will vary between 95.8 (equation 1) in a perfectly concentrated system in which all the production occurs in one month (e.g. paddy only, on a farm) on the one hand, and 0 (equation 2) on the other hand with a perfectly dispersed income (equal income every month of the year) (McConnell and Dharampala, 1973).

Equation 1:

$$95.8 = \sqrt{(100-8.3)_1^2 + (8.3-0)_2^2 + (8.3-0)_3^2 + ...(8.3-0)_{12}^2}$$
$$= \sqrt{9,167}$$

Equation 2:

$$0 = \sqrt{(8.3-8.3)_1^2 + (8.3-8.3)_2^2 + (8.3-8.3)_3^2 + ...(8.3-8.3)_{12}^2}$$

where 8.3 = % income every month in a cycle of 12 months

A survey of 30 home gardens in Sri Lanka indicated that 11 farms had a time concentration index of less than 50, and 19 farms had an index of more than 50. The index was identified to be associated with the clove crop; of the 11 farms which had a low index, only one grew cloves; all the 19 farms which had an index value higher than 50, grew cloves. When cloves were excluded, the time concentration index became low, the mean value then being only 32, and as many as 25 farms (out of 30) had index values of less than 50 (McConnell and Dharampala, 1973).

In a case study conducted on a 0.12 ha plot of undeveloped land in Trivandrum district in the southern zone of Kerala, a farmer undertook a three-tier planting of fodder, cassava, vegetables, banana, clove and coconut with *Leucaena* being planted all round. There were 23 coconut, 12 clove, 56 banana, 49 pineapple and 30 pepper plants; fodder grasses were planted in a 0.03 ha plot. One cow was maintained. Chemical fertilisers were not used. The net income derived during the first three years (when the cash crops had not started yielding) was Rs. 1,550, 3,848, and 3,950 in the first, second and third years, respectively (KGSN, 1985). The expected net income from cash crops at the 1986 prices, when these crops were to start full bearing was estimated to be: cloves, Rs. 3,800; coconut, Rs. 900, other plants, Rs. 1,500, and livestock, Rs. 3,000, all adding up to a total of Rs. 9,200 from a 0.12 ha plot.

In the Pokkali zone in Kerala, where paddy is cultivated in stagnant water of up to one-metre depth, prawn culture is successfully practised; only one crop of paddy is taken in these fields, and the stubbles of paddy are used as feed for prawns. One hectare of land can produce 2,000 kg paddy and an estimated 750 kg prawn, giving a total return of Rs. 11,500 on an investment of Rs. 4,990 (excluding the family labour) per annum (Mathew, 1985; SPB, 1985).

It may be noted that although these income values are in the realm of speculation, due to experiences elsewhere outside India, one fact is indisputable that the practice is remunerative and would provide a good subsistence (some one has called it affluent subsistence). No benefit : cost ratio values however, are available.

The multitude of crop species in the home gardens provide food, fuelwood, fodder, fruit, fertiliser, beverages, spices, small timber, regular cash inflow, and work to the farmer with low risk on the one hand, and conserve the production base on the other. The use of waste materials for feeding cattle, pigs, poultry and fish results in an efficient recycling of the wastes.

Constraints and Potential

In spite of all the above benefits, the major constraint of the practice is that it is the least understood scientifically. There is tremendous scope of improvement of the practice. One may think that the practice is working at random. It cannot be so. The farmers obviously understand the biophysical, social and economic advantages of the practice. The improvement of the practice is very challenging, potentially more promising due to the availability of new short-duration and high-yielding varieties of many crops. Some of the research components, which call for understanding are: spatial arrangements of the various crop varieties, their compatibility or otherwise, use of improved, high-yielding, fast-growing varieties, input : output relationships, long-term effects on stability of the production base, and protective benefits.

Nair and Sreedharan (1986) have identified the following research needs: water and nutrient requirements of crops in a polycropping system; rooting pattern and root distribution studies of the component crops for understanding nutrient-moisture interaction and utilisation; spatial and temporal arrangement of crops for the efficient conversion of solar energy; identifying crop associations which can be fitted into different intensities of shade; integrated production approach to be worked out involving crops, livestock, microbes, fertilisers, pests; extrapolability of one system into similar agro-ecological zones and working out an economic analysis; complementary and competitive interaction effects between the crop components.

The practice of home gardens has a great scope in the tropical high rainfall agroecological conditions in Tamil Nadu, Karnataka, Orissa, West Bengal and in the entire northeastern hill region.

8.2 ANIMALS IN SHIFTING CULTIVATION

It has already been stated that shifting cultivators maintain pigs, livestock and poultry. Farming systems research (Anonymous, 1983) has indicated that all the farming systems have some area available which yield fodder and byproducts for use by animals. For example, in shifting cultivation, paddy straw and tapioca leaf meals are available from the area under cultivation; in the case of the agrohorticulture system, paddy straw is available from bench terrace risers, pineapple waste and cowpea are available from the horticulture area, and cowpea fodder and sweet potato leaf meal are available from the intercrops in the horticultural area; in the agricultural system, paddy straw and pulse straw are available from bench terrace surfaces and legume fodder from bench terrace risers. The yield of straw and fodder has been evaluated in terms of dry matter, digestible protein and nutrients, and energy (Table 42).

Table 42: Production of straw and fodder from alternative farming systems

Farming system	Straw and fodder	Area of availability	Per cent of total area
Shifting cultivation	a. Paddy straw (*kharif*)	Surface area	100.0
	b. Tapioca leaf meal (*rabi*)	Surface area	100.0
Agrohorti-culture	a. Paddy straw (*kharif*)	Terrace surface	33.0
	b. *Stylosanthes guyanensis* (perennial)	Terrace risers	8.4
	c. Pineapple waste	Horticultural area	67.0
	d. Cowpea fodder (*kharif*)	Interspaces of horticultural area	20.0
	e. Sweet potato leaf meal (*rabi*)	Interspaces of horticultural area	20.0
Agriculture (Paddy-pulse)	a. Paddy straw (*kharif*)	Terrace surface	100.0
	b. *Stylosanthes guyanensis* (perennial)	Terrace risers	30.4
	c. Black gram straw (*rabi*)	Terrace surface	100.00
Agriculture (Paddy, maize-pulse)	a. Paddy and maize straw (*kharif*)	Terrace surface	64.0
	b. *Stylosanthes guyanensis*	Terrace risers	36.0
	c. Black gram straw (*rabi*)	Terrace surface	64.0

Source: Anonymous, 1983

Based on the above parameters, it has been concluded that shifting cultivation, agrohorticultural practice, agriculture (paddy and pulse), and agriculture (paddy, maize and pulse) could support 0.50-0.78, 1.18-2.73, 1.05-3.26 and 1.07-1.47 animals, respectively (Anonymous, 1983). It would appear that dry matter produced under any farming system could support potentially more animals than any of the other four parameters—digestible crude protein, total digestible nutrients, digestible energy and metabolisable energy.

APPENDICES

APPENDIX 1: Research Institutes and Agricultural Universities Where Agroforestry Research is Carried Out

Research Institutes

1.* Central Soil & Water Conservation Research & Training Institute (CSWCRTI), Dehradun (U.P.) and its Research Centres at:

1a. Chandigarh; 1b. Agra; 1c. Kota; 1d. Vasad; 1e. Hyderabad; 1f. Bellary; 1g. Ootacumand; 1h. Datia; 1j. Koraput.

2.* Central Arid Zone Research Institute (CAZRI), Jodhpur (Rajasthan)

3. Central Sheep and Wool Research Institute (CSWRI), Avikanagar (Rajasthan)

4. Central Institute for Research on Goats (CIRG), Makhdoom (U.P.)

5.* Indian Grassland and Fodder Research Institute, (IGFRI), Jhansi (U.P.)***

6. Indian Lac Research Institute, (ILRI), Ranchi (Bihar)

7. Central Inland Fisheries Research Institute, (CIFRI), Barrackpore (W.B.)

8.* ICAR Research Complex for NEH Region, Shillong (Meghalaya) and its Research, Centres at:

8a. Kohima; 8b.* Imphal; 8c. Aizwal; 8d. Agartala, 8e.* Gangtok

9. Indian Institute of Horticulture Research (IIHR), Bangalore (Karnataka)

10. Central Palantation Crops Research Institute (CPCRI), Kasargod (Kerala), and its Research Centres at:

10a. Vittal; 10b. Kannara; 10c. Palode; 10d. Hirehalli; 10e. Kidu; 10f. Kahikuchi; 10g. Mohitnagar.

11. Central Agricultural Research Institute for Andaman & Nicobar Group of Islands, Port Blair (A & N)

12. Forest Research Institute and Colleges (FRI&C), Dehradun with its Regional Centres, Dehradun (U.P.)

13a. Tea Research Institute (TRI), Tocklai (Assam)

13b. Tea Research Department of UPASI, Ootacamand (T.N.)

(contd.)

Agricultural Universities

A1.**	Sher-e-Kashmir University of Agricultural Sciences & Technology, (SKUAST), Srinagar, (J&K) (Summer);
A2.	Sher-e-Kashmir University of Agricultural Sciences & Technology, (SKUAST), Jammu, (J&K) (Winter)
B1.	Himachal Pradesh Krishi Vishwa Vidyalaya (HPKVV), Palampur
B2.**	Y.S. Parmer University of Horticulture & Forestry (YSPUHF), Solan (HP)
C.**	Haryana Agricultural University (HAU), Hissar (Haryana)
D.**	Punjab Agricultural University (PAU), Ludhiana (Punjab)
E1.**	G.B. Pant University of Agriculture & Technology (GBPUAT), Pantnagar (U.P.)
E2.	G.B. Pant University of Agriculture & Technology (GBPUAT), Ranichauri Campus (U.P.)
F.**	C.S. Azad University of Agriculture & Technology (CSAUAT), Kanpur (U.P.)
G.**	Narendra Deva University of Agriculture & Technology (NDUAT), Faizabad (U.P.)
H.**	Rajendra Agricultural University (RAU), Pusa (Bihar)
J.**	Birsa Agriculture University (BAU), Ranchi (Bihar)
K.**	Bidhan Chandra Krishi Vishwa Vidyalaya (BCKVV), Kalyani (WB)
L.**	Assam Agricultural University (AAU), Jorhat (Assam)
M.**	M.L. Sukhadia University (MLSU), Udaipur (Rajasthan)
N.**	Gujarat Agricultural University (GAU), Dantiwada (Gujarat)
P.**	Jawahar Lal Nehru Krishi Vishwa Vidyalaya (JNKVV), Jabbalpur (M.P.)
Q.**	Punjabrao Krishi Vidyapeeth (PKV), Akola at Agricultural College, Nagpur (Maharashtra)
R.**	Mahatma Phule Krishi Vidyapeeth (MPKV), Rahuri (Maharashtra)
S.**	Marathwada Agricultural University (MAU), Parbhani (Maharashtra)
T.**	Konkan Krishi Vidyapeeth (KKV), Dapoli (Maharashtra)
U.**	Orissa University of Agriculture & Technology (OUAT), Bhubaneswar (Orissa)
V.**	Andhra Pradesh Agricultural University (APAU), Rajendranagar (A.P.)
W.**	University of Agricultural Sciences (UAS), Dharwad Campus (Karnataka)
X.**	Tamil Nadu Agricultural University (TNAU), Coimbatore (T.N.)
Y.**	Kerala Agricultural University (KAU), Mannuthy (Kerala)

*All the Institutes carry out research in agroforestry as a part of their charter. Those marked with an asteric are also centres of All India Coordinated Research Project on Agroforestry.
**Universities where centres of All India Coordinated Research Project on Agroforestry are located.
***National Research Centre for Agroforestry is also located at the Campus of IGFRI.

APPENDIX 2: Important Multipurpose Tree Species

Botanical Name	Climatic zone	Planting method*	Important uses
Acacia auriculiformis	Dry to moist tropical	DS, EP	Fuelwood, Soil conservation
Acacia catechu	Dry to moist tropical	DS, EP, ST	Catechu, Soil conservation
Acacia farnesiana	Dry to moist tropical	DS, EP	Gum, Fuelwood
Acacia leucophloea	Dry to moist tropical	DS, EP	Timber turning, Fodder,
Acacia mearsnii	Temperate	DS, EP, RS	Timber, Fuelwood
Acacia modesta	Dry to moist tropical	DS, EP	Timber, Soil conservation, Fodder, Fuelwood
Acacia nilotica ssp. *cupressiformis*	Dry tropical	DS, EP	Small timber poles, Fuelwood, Fodder, Gum, Tannin, Soil conservation
Acacia nilotica ssp. *indica*	Dry to moist tropical	DS, EP	Fodder, Soil conservation, Agricultural implements, Tannin, Fuelwood
Acacia tortilis	Dry tropical	EP	Fodder, Fuelwood
Aegle marmelos	Dry to moist tropical	DS, EP, Air layering	Fruit, Pickles, Medicinal
Aesculus indica	Moist tropical to sub-tropical	EP, ST	Fodder, Timber, Food
Ailanthus excelsa	Dry to moist tropical	DS, EP, ST, BC	Fodder, Soil conservation
Albizia amara	Dry tropical	DS, EP	Timber, Fuelwood
Albizia falcataria	Tropical moist	DS, EP, ST	Soil conservation, Fuelwood
Albizia lebbek	Very dry to wet tropical	DS, EP, ST	Fodder, Oilseed, Timber, Soil conservation, Fuelwood
Albizia odoratissima	Moist tropical	DS, EP, ST	Fodder, Timber
Albizia stipulata	Tropical moist to sub-tropical	DS, EP, ST	Timber, Fodder, Fuelwood
Alnus nepalensis	Temperate to sub-tropical	DS, EP	Small timber poles, Tannin, Ornamental, Soil conservation, Medicinal
Alnus nitida	Moist temperate	SL, DS	Fodder, Soil conservation, Fuelwood
Anacardium occidentale	Dry to moist tropical	DS, ST, Air layering	Fuelwood, Soil conservation
Anthocephalus chinensis	Moist to wet tropical	EP, ST	Ornamental, Fodder, Fruit, Pulpwood, Fuelwood
Artocarpus heterophyllus	Moist tropical	DS, EP (in polythene bags)	Fruit, Timber, Fuelwood
Artocarpus lokoocha	Tropical moist	DS, EP	Edible fruit, Industrial timber, Fuelwood

(contd.)

Appendix 2 contd.

Botanical Name	Climatic zone	Planting method*	Important uses
Azadirachta indica	Very dry to moist tropical	DS, EP, RS	Fodder, Soil conservation, Timber, Medicinal, Oil, Fuelwood
Bassia butyracea	Temperate	—	Food, Fuelwood, Fodder, Soil conservation
Bauhinia purpurea	Moist tropical	DS, EP, ST	Tannin, Fodder, Agricultural implements
Bauhinia racemosa	Tropical to sub-tropical	DS, EP, ST	Fodder, Fuelwood
Bauhinia roxburghiana	Tropical to sub-tropical	BC, layering, budding or grafting	Fodder, Gum, Medicinal
Bauhinia vahlii	Temperate	—	Fuelwood, Fodder
Bauhinia variegata	Moist tropical to sub-tropical	DS, EP, ST	Fodder, Food, Timber
Bombax ceiba	Moist tropical to sub-tropical	DS, EP, ST, BC	Soil conservation, Match sticks, Packing cases, Plywood
Carpinus viminea	Temperate	—	Fodder
Cassia fistula	Tropical moist	DS, EP, ST	Medicinal, Fuelwood
Cassia siamea	Dry to moist tropical	DS, EP, ST	Soil conservation, Fodder, Fuelwood
Casuarina equisetifolia	Moist tropical	EP	Timber, Fuelwood
Celtis australis	Moist tropical	DS, EP	Soil conservation, Fuelwood, Tool handles, Timber
Cordia myxa	Temperate	—	Fodder
Cornus capitata	Temperate	—	Fruit
Corylus colaurna	Temperate	—	Fruit, Oilseed, Fodder
Cupressus cashmeriana	Temperate to sub-tropical	EP	Small timber poles, Ornamental, Soil conservation, Tubers
Cupressus torulosa	Temperate to sub-tropical	EP	Industrial timber, Small timber poles, Oil, Ornamental, Soil conservation
Dalbergia latifolia	Tropical moist	DS, EP, ST	Fodder, Timber, Fuelwood
Dalbergia sissoo	Dry to moist tropical	DS, EP, ST, BC, RS	Timber, Soil conservation, Fodder, Fuelwood
Dendrocalamus strictus	Dry tropical to moist tropical	DS, EP	Industrial timber, Small timber poles, Fodder, Edible shoots, Gum, Soil conservation, Medicinal

Derris indica	Dry to moist tropical	DS, EP, ST	Oil seed, Soil conservation, Fuelwood
Emblica officinalis	Very dry to wet tropical	EP, DS, Budding	Fruit, Fuelwood
Erythrina suberosa	Moist tropical	RC, RS, ST, BC	Soil conservation, Fuelwood
Erythrina variegata	Moist tropical	EP, ST, BC	Timber, Fuelwood
Eucalyptus camaldulensis	Dry to moist tropical	EP	Industrial timber, Oil, Gum, Tannin, Ornamental, Soil conservation
Eucalyptus globulus	Temperate	EP	Timber, Pulpwood, Fuelwood
Eucalyptus tereticornis	Dry to moist tropical	EP	Pulpwood, Fuelwood
Ficus auriculata syn. *F. roxburghii*	Moist tropical	EP, BC	Fruit, Fodder, Fuelwood,
Ficus bengalensis			
Ficus nemoralis			
Ficus religiosa			
Ficus rigida syn. *F. glaberrima*			
Ficus semicordata			
Ficus subincisa syn. *F. clavata*			
Fraxinus spp.	Sub-tropical to temperate	EP	Soil conservation, Fodder, Fuelwood, Timber
Grewia optiva	Sub-tropical to temperate	EP, ST, BC	Fodder, Fuelwood
Grevillea robusta	Sub-tropical	DS, EP	Fuel, Timber, Packing cases
Hardwickia binata	Dry tropical	DS, EP	Fodder, Fuelwood, Agricultural implements
Holoptelea integrifolia	Moist tropical	DS, EP, ST	Fodder, Timber, Fuelwood
Juglans regia	Temperate	DS, EP	Fruit, Carving and ornamnetal timber, Gun butts
Kydia calycina	Moist tropical	DS, EP, ST, tree stumping	Fodder
Lagerstroemia spp.	Tropical moist	SL, ST	Industrial timber, Small timber poles, Fuelwood, Fodder, Gum, Tannin, Ornamental, Soil conservation
Lannea coromandelica	Dry to moist tropical	DS, BC, ST	Soil conservation, Fence posts
Leucaena leucocephala	Dry to moist tropical	DS, SL	Fodder, Soil conservation
Madhuca longifolia	Dry to moist tropical	DS, EP, ST	Oil seed, Fruit, Timber, Fuelwood
Mangifera indica	Moist tropical to sub-tropical	DS, EP, Grafting	Fruit, Timber, Fuelwood, Packing cases, Plywood, Fodder
Melia azedarach	Moist tropical to sub-tropical	DS, EP, ST, BC, RS	Fodder, Timber

(contd.)

Appendix 2 contd.

Botanical Name	Climatic zone	Planting method*	Important uses
Mesua ferrea	Tropical moist	DS, EP	Industrial timber, Small timber poles, Fuelwood, Fruit, Oil seed, Gum, Tannin, Ornamental, Soil conservation, Medicinal
Michella doltsopa	Tropical moist	EP	Industrial timber, Small timber poles, Ornamental, Soil conservation, Medicinal
Michella velutina	Sub-tropical	EP	Industrial timber, Small timber poles, Ornamental, Soil conservation, Medicinal
Moringa oleifera	Moist tropical	DS, BC	Fruit, Fodder
Morus alba	Moist tropical to sub-tropical	BC	Timber, Fuelwood
Morus macroura	Sub-tropical	EP, ST, BC	Industrial timber, Small timber poles, Fodder, Ornamental, Soil conservation, Fruit
Morus serrata	Sub-tropical	EP, ST, BC	Industrial timber, Small timber poles, Fodder, Ornamental, Soil conservation
Myrica nagi	Temperate	—	Fruit, Medicinal
Olea ferruginea	Sub-tropical to temperate	EP	Fodder, Fruit, Oil seed
Olea glandulifera	Temperate	—	Fuelwood, Fodder, Timber
Ougeinia dalbergioides	Temperate to sub-tropical	—	Fuelwood, Fodder, Timber
Parkinsonia aculeata	Very dry to dry tropical	DS	Fuelwood
Peltophorum pterocarpum	Dry to moist tropical	DS, BC, EP	Timber, Tannin, Medicinal, Fodder, Fuelwood
Pinus roxburghii	Sub-tropical	DS, EP	Industrial timber, Small timber poles, Fodder, Gum, Tannin, Oil, Ornamental, Soil conservation
Pithecellobium dulce	Tropical	DS, EP, BC	Tannin, Dye, Medicinal, Timber
Platanus orientalis	Temperate	EP	Small timber, Ornamental, Soil conservation, Medicinal
Populus ciliata; Populus deltoides	Sub-tropical to temperate	SC	Soil conservation, Fodder, Packing cases, Pulpwood, Fuelwood
Prosopis cineraria	Dry tropical	DS, EP	Fodder, Fuelwood

Prosopis juliflora	Very dry to moist tropical	DS, EP	Soil conservation, Fodder, Fuelwood
Prunus armeniaca	Temperate	EP	Small timber poles, Fuelwood, Fruits, Ornamental, Soil conservation
Prunus cerasoides	Temperate	EP	Fruit, Oil, Gum, Tannin, Ornamental, Soil conservation, Medicinal
Prunus persica	Temperate	EP	Small timber poles, Fuelwood, Fruit, Oil, Gum, Tannin, Ornamental, Soil conservation, Medicinal
Prunus spp.	Sub-tropical to temperate	EP, DS	Fruit
Psidium guajava	Tropical to sub-tropical	EP, BC	Fruit
Pterocarpus marsupium	Moist tropical	EP, ST, DS	Fodder, Timber, Medicinal, Fuelwood
Pyracanth crenulata	Temperate	—	Fruit
Pyrus pashia	Temperate	—	Fruit, Small timber, Fodder
Quercus floribunda *Quercus glauca* *Quercus leucotrichophora* *Quercus semecarpifolia* *Quercus* spp.	Temperate	DS, EP	Industrial timber, Small timber poles, Fuelwood, Fodder, Tannin, Ornamental, Soil conservation
Robinia pseudacacia	Moist Temperate	DS, EP	Fodder, Soil conservation, Fuelwood
Salix spp.	Temperate	Pollards, BC, by clonal propagation	Soil conservation, Match wood, Sports goods, Fodder, Fuelwood
Santalum album	Dry tropical	DS, EP, RS	Industrial timber, Oil, Ornamental, Soil conservation, Medicinal
Sapindus mukorossi	Tropical	DS, EP	Soap nut for washing, Fuelwood
Saraca asoca	Tropical, sub-tropical and dry tropical	DS, EP	Small timber poles, Fodder, Ornamental, Soil conservation, Medicinal
Schleichera oleosa	Tropical	RS	Industrial timber, Small timber poles, Fuelwood, Fruit, Oil, Gum, Tannin, Ornamental Soil conservation, Medicinal
Sesbania grandiflora	Tropical moist to sub-tropical	DS, EP	Fence posts, Tannin, Gum, Medicinal, Fodder, Fuelwood, Pulp (fibre)
Shorea robusta	Tropical moist	DS, EP, ST	Industrial timber, Small timber poles, Fuelwood, Oil, Gum, Tannin, Ornamental, Soil conservation, Medicinal

(contd.)

Appendix 2 contd.

Botanical Name	Climatic zone	Planting method*	Important uses
Swietenia mahagoni	Sub-tropical to moist tropical	DS, EP, ST	Industrial timber, Small timber poles, Gum, Tannin, Ornamental, Soil conservation
Symplocos crataegoides	Temperate	—	Fodder
Syzygium cuminii	Moist tropical	DS, EP, ST	Fruit, Timber, Fuelwood
Tamarindus indica	Dry to moist tropical	DS, EP	Fruit, Timber, Fuelwood
Tectona grandis	Dry and moist tropical	ST, EP	Industrial timber, Small timber poles, Fuelwood, Fodder, Tannin, Ornamental, Medicinal
Terminalia alata	Tropical	DS, EP, ST	Industrial timber, Small timber poles, Fuelwood, Fruit, Gum, Tannin, Ornamental, Soil Conservation, Medicinal
Terminalia arjuna	Tropical	DS, EP, ST	Industrial timber, Small timber poles, Fuelwood, Fodder, Tannin, Ornamental, Soil conservation, Medicinal
Terminalia bialata	Tropical moist	DS, EP, ST	Industrial timber, Small timber poles, Gum, Tannin, Ornamental, Soil conservation, Medicinal
Terminalia chebula	Tropical moist	DS, EP, ST	Industrial timber, Small timber poles, Gum, Tannin, Ornamental, Soil conservation, Medicinal
Terminalia myriocarpa	Tropical moist	DS, EP, ST	Industrial timber, Small timber, Fuelwood, Tannin, Ornamental, Soil conservation, Medicinal
Toona ciliata syn. *Cedralla toona*	Temperate	—	Fuelwood, Fodder
Zizyphus mauritiana	Sub-tropical	EP, DS	Industrial timber, Fodder, Fruit, Tannin, Ornamental, Soil conservation, Medicinal

Source: FRI&C

HAPPRC (All the temperate region trees without planting method)

*Note : DS = Direct sowing, RS = Root suckers, EP = Entire planting, RC = Root cutting, ST = Stump planting, BC = Branch planting

APPENDIX 3: Chemical Composition, Digestibility Coefficients and Nutritive Value of Tree Leaf Fodders, Shrubs and Grasses

Name	Chemical composition on DM basis					Digestibility coefficients				Nutritive value					
										kg/100 kg dry matter			Mcal/kg dry matter		NR
	CP	CF	NFE	EE	TA	CP	EE	CF	NFE	DCP	TDN	NR	DE	ME	
Trees Leaves and Shrubs															
Acacia catechu	13.0	22.6	51.0	4.6	9.8	24	60	31	60	2.9	46.3	33.3	2.0	1.7	15.0
Acacia nilotica (syn. *A. arabica*), average	11.0	26.7	51.4	3.8	7.1	–	–	–	–	–	–	–	–	–	–
*Acacia senegal**	10.3	9.7	65.7	–	–	–	–	–	–	–	–	–	–	–	–
Aegle marmelos	15.1	16.4	52.8	1.5	14.1	71	62	30	73	10.8	56.7	47.3	2.5	2.1	4.3
Ailanthus excelsa	19.9	12.7	51.8	3.5	–	–	–	–	–	–	–	–	–	–	–
Albizia lebbek, average	21.7	31.4	36.1	3.7	9.3	67	9	41	68	11.6	49.3	31.0	2.2	1.8	3.3
*Albizia stipulata***	–	–	–	–	–	–	–	–	–	4.87	40.23	–	1.77	1.44	–
Allmania nudiflora	10.3	25.8	55.6	2.8	5.5	–	–	–	–	–	–	–	–	–	–
Amorphophallus paeniifolius (syn. *A. campannulatus*)	15.3	14.2	50.9	3.3	16.2	–	–	–	–	–	–	–	–	–	–
Anogeissus latifolia	7.4	24.2	55.4	3.6	9.5	–	–	–	–	–	–	–	–	–	–
Anogeissus pendula	7.60	19.0	65.3	–	–	–	–	–	–	–	–	–	–	–	–
Anogeissus sericea var. *nummularia* (syn. *A. rotundifolia*)	8.90	20.0	62.2	–	–	–	–	–	–	–	–	–	–	–	–
Artocarpus heterophyllus (syn. *A. integrifolia*), average	12.8	19.9	52.3	2.9	11.7	–	–	–	–	–	–	–	–	–	–
Azadirachta indica, average	17.4	17.5	51.3	3.8	10.0	52	58	23	68	8.4	53.3	41.3	2.4	1.9	5.4
Bauhinia racemosa	15.4	18.5	54.6	3.3	–	–	–	–	–	–	–	–	–	–	–

(contd.)

Appendix 3 contd.

Name	Chemical composition on DM basis					Digestibility coefficients				Nutritive value					
	CP	CF	NFE	EE	TA	CP	EE	CF	NFE	kg/100 kg dry matter			Mcal/kg dry matter		NR
										DCP	TDN	NR	DE	ME	
Bauhinia variegata, average	15.7	31.9	40.8	1.8	9.7	58	42	64	60	9.2	55.5	37.0	2.4	2.0	5.1
Calligonum polygonoides	7.4	12.5	67.3	–	–	–	–	–	–	–	–	–	–	–	–
Careya arborea	10.4	25.9	42.6	7.7	7.5	1	18	66	53	0.2	43.1	28.0	1.9	1.6	86.2
Cassia auriculata	10.9	12.5	67.3	–	–	–	–	–	–	–	–	–	–	–	–
Celtis tetrandra	14.0	16.6	44.0	4.1	–	–	–	–	–	–	–	–	–	–	–
Cordia dichotoma (syn. *C. obliqua*)	15.1	16.4	52.8	1.5	14.1	71	30	58	76	5.1	26.9	12.7	1.2	1.0	4.2
Dalbergia sissoo, average	9.6	27.7	48.8	3.2	10.8	56	27	30	72	9.1	52.2	38.0	2.3	1.9	4.7
Delonix elata	25.6	7.2	46.5	7.5	13.2	–	–	–	–	–	–	–	–	–	–
Dendrocalamus strictus, average	14.6	25.6	43.1	1.6	15.2	66	33	58	50	9.3	48.9	32.7	2.2	1.8	4.2
Dichrostachys nutan	–	–	–	–	–	–	–	–	–	5.9	61.4	–	2.7	2.2	–
Diospyros tomentosa	7.1	25.3	56.0	2.2	9.4	0	0	28	48	0	34.1	19.5	1.5	1.2	–
Ehretia laevis	13.5	18.0	51.2	6.0	11.4	62	18	36	73	8.4	54.9	44.5	2.4	2.0	5.5
Ficus auriculata (syn. *F. roxburghii*)	13.4	7.7	64.9	4.7	–	–	–	–	–	–	–	–	–	–	–
Ficus bengalensis, average	9.8	21.9	53.0	2.7	12.5	21	29	26	65	2.0	44.5	29.0	2.0	1.6	21.4
Ficus racemosa (syn. *F. glomerata*)	11.2	12.3	59.0	2.4	15.1	60	48	63	62	6.6	53.8	47.0	2.4	2.0	7.0
Ficus religiosa	9.7	27.0	45.8	2.7	14.9	59	36	22	52	7.0	38.3	27.7	1.7	1.4	4.5
Ficus rumphii	9.6	28.3	49.9	3.4	–	–	–	–	–	–	–	–	–	–	–

Ficus virens	–	–	–	–	–	56	62	32	58	5.4	47.1	32.5	2.1	1.7	7.7
(syn. *F. lacor*)															
Young	11.2	27.7	51.3	2.2	7.7	56	53	62	68	6.3	62.0	46.0	2.7	2.2	8.9
Mature	12.5	20.0	49.9	3.5	13.1	32	24	40	53	4.0	40.7	24.0	1.8	1.5	9.2
Autumn	7.3	25.5	50.2	2.7	14.2	6	30	33	43	0.4	32.2	17.5	1.4	1.2	75.5
Grewia elastica	18.1	12.0	46.2	11.0	–	–	–	–	–	–	–	–	–	–	–
Grewia optiva															
(syn. *G. oppositifolia*)															
Maximum	24.9	–	–	–	–	–	–	–	–	–	–	–	–	–	–
Minimum	13.3	–	–	–	–	–	–	–	–	–	–	–	–	–	–
Average	20.2	–	–	–	–	–	–	–	–	–	–	–	–	–	–
July lopped	–	–	–	–	–	80	36	54	70	16.7	54.6	47.1	2.4	2.0	2.3
September lopped	–	–	–	–	–	78	37	47	75	15.8	60.2	53.6	2.6	2.2	2.8
Grewia tenax	10.3	18.9	61.1	–	–	–	–	–	–	–	–	–	–	–	–
Haldina cordifolia	9.5	14.0	64.0	4.8	7.6	28	20	23	67	2.7	50.9	42.5	2.2	1.8	18.1
Helicteres isora	13.2	19.8	53.0	3.0	10.9	73	23	41	74	9.7	58.4	46.8	2.6	2.1	5.0
Hevea brasiliensis	13.0	32.2	43.1	4.4	7.3	–	–	–	–	–	–	–	–	–	–
Holoptelea integrifolia	13.7	–	–	1.9	14.5	68	26	–	–	5.2	58.1	–	2.6	2.1	10.2
Ipomoea spp.	12.5	12.2	55.8	2.2	17.4	–	–	–	–	–	–	–	–	–	–
Kydia calycina	12.5	23.7	46.1	3.3	14.4	63	49	47	49	7.9	45.2	31.5	2.0	1.6	4.7
Lagerstroemia parviflora	7.8	17.3	58.2	5.7	11.1	11	31	44	64	0.9	49.5	39.5	2.2	1.8	56.6
Lannea coromandelica	11.4	16.1	59.1	4.2	9.2	43	62	23	69	4.9	55.2	35.0	2.4	2.0	10.3
(syn. *L. grandis*)															
Leucaena leucocephala	16.7	12.6	51.1	7.1	12.5	–	–	–	–	–	–	–	–	–	–
(syn. *L. glauca*)															
Madhuca latifolia, average	9.5	19.5	60.0	3.9	7.3	71	62	30	73	0.0	37.0	25.3	1.6	1.3	–
Madhuca longifolia	–	–	–	–	–	–	–	–	–	7.0	37.0	–	1.6	1.3	–
Mallotus philippinensis	13.4	29.7	44.6	3.6	8.7	59	24	31	62	7.9	46.6	29.3	2.1	1.7	4.9

(contd.)

Appendix 3 contd.

Name	Chemical composition on DM basis					Digestibility coefficients				Nutritive value					
	CP	CF	NFE	EE	TA	CP	EE	CF	NFE	kg/100 kg dry matter			Mcal/kg dry matter		NR
										DCP	TDN	NR	DE	ME	
Mangifera indica	9.3	23.7	52.6	2.6	11.8	–	–	–	–	–	–	–	–	–	–
Manilkara hexandra (syn. *Mimusops hexandra*)	9.4	23.1	55.0	6.1	7.4	–	–	–	–	–	–	–	–	–	–
Maytenus senegalensis (syn. *M. emarginata*)	9.6	11.5	55.2	2.5	–	–	–	–	–	–	–	–	–	–	–
Millettia extensa (syn. *M. auriculata*)	22.7	32.5	30.9	4.6	9.3	69	36	32	49	15.5	44.9	26.0	2.0	1.6	1.9
Mitragyna parviflora	7.7	19.6	60.7	3.3	3.7	19	0	31	70	1.5	49.9	38.5	2.2	1.8	33.0
Morinda pubescens (syn. *M. tinctoria*)	15.1	22.7	47.8	2.9	11.5	–	–	–	–	–	–	–	–	–	–
Moringa oleifera	15.6	17.9	48.7	4.4	13.4	71	30	58	76	11.1	61.5	50.8	2.7	2.2	4.5
Morus alba	15.0	15.3	48.0	7.4	14.3	71	4	54	83	10.7	59.6	50.8	2.6	2.1	4.6
Morus serrata	14.0	17.8	43.4	1.9	–	–	–	–	–	–	–	–	–	–	–
Ougeinia dalbergioides	11.6	26.5	47.6	4.3	10.0	32	42	36	62	3.7	45.6	30.2	2.0	1.6	11.2
Ougeinia oojeinensis	–	–	–	–	–	–	–	–	–	4.3	45.5	–	2.0	1.6	–
Musa spp.															
Leaves	9.5	23.1	48.5	5.6	13.3	–	–	–	–	–	–	–	–	–	–
Stem	2.8	13.8	66.7	1.2	15.6	–	–	–	–	–	–	–	–	–	–
Bulb	8.7	24.2	47.2	2.8	17.1	–	–	–	–	–	–	–	–	–	–
Phoenix dactylifera	11.6	27.7	51.1	1.6	8.0	–	–	–	–	–	–	–	–	–	–
Prosopis cineraria	15.3	17.5	54.1	3.2	10.0	–	–	–	–	–	–	–	–	–	–
Prosopis juliflora	21.4	20.8	50.0	–	–	–	–	–	–	–	–	–	–	–	–

Quercus floribunda	9.6	29.1	51.8	4.5	5.1	44	4	25	61	4.2	43.2	26.3	1.9	1.6	2.3
(syn. *Q. dilatata*)															
Quercus glauca	9.6	29.0	49.6	4.1	7.6	48	15	25	54	4.6	39.8	23.2	1.8	1.4	7.6
Quercus leucotrichophora	10.2	31.3	48.4	4.8	5.2	57	29	29	54	5.8	43.8	25.8	1.9	1.6	6.6
(syn. *Q. incana*)															
Salvadora oleoides	9.6	9.3	40.2	–	–	–	–	–	–	–	–	–	–	–	–
Salvadora persica	14.2	9.4	44.9	–	–	–	–	–	–	–	–	–	–	–	–
Schleichera oleosa	10.4	32.3	49.2	1.9	6.2	33	34	31	68	13.4	47.5	28.8	2.1	1.7	13.0
Sesbania grandiflora	37.3	6.4	41.0	5.5	9.8	–	–	–	–	–	–	–	–	–	–
Shorea robusta	10.1	27.4	55.4	3.2	3.9	11	36	24	58	1.1	42.7	27.0	1.9	1.5	36.5
Syzygium cuminii	8.8	19.8	62.5	3.1	5.8	1	33	70	45	0.1	43.8	34.5	1.9	1.6	75.6
(syn. *Eugenia jambolana*)															
Tamarindus indica															
Monsoon lopped	11.2	18.7	55.4	5.9	8.8	–	–	–	–	–	–	–	–	–	–
Winter lopped	12.9	14.2	51.9	8.1	13.1	–	–	–	–	–	–	–	–	–	–
Summer lopped	15.2	21.4	48.5	7.2	7.8	–	–	–	–	–	–	–	–	–	–
Tecomella undulata	12.2	15.8	63.0	–	–	–	–	–	–	–	–	–	–	–	–
Terminalia alata	8.9	21.8	54.1	4.9	10.3	0	35	27	47	0.0	34.9	22.3	1.5	1.3	–
(syn. *T. tomentosa*)															
Terminalia bellerica	8.6	18.6	60.1	4.7	8.0	10	22	56	68	0.9	54.5	43.3	2.4	2.0	62.3
Tinospora cordifolia	11.2	17.5	61.2	2.5	7.8	–	–	–	–	–	–	–	–	–	–
Tribullus lanuginosus	18.5	37.0	23.9	2.9	–	–	–	–	–	–	–	–	–	–	–
(syn. *T. terrestris*)															
Ziziphyus mauritiana	8.6	30.1	48.8	1.7	10.7	36	62	27	34	3.1	30.6	15.8	1.3	1.1	9.0
Mature Average															
of all seasons	15.4	14.2	56.7	2.7	10.2	–	–	–	–	–	–	–	–	–	–
Zizyphus nummularia	11.5	33.8	46.8	1.6	6.2	47	28	54	69	5.4	51.1	33.0	2.3	1.9	9.5

(contd.)

Appendix 3 contd.

Name	Chemical composition on DM basis					Digestibility coefficients				Nutritive value					
	CP	CF	NFE	EE	TA	CP	EE	CF	NFE	kg/100 kg dry matter			Mcal/kg dry matter		NR
										DCP	TDN	NR	DE	ME	
Hays															
Apluda mutica															
(syn. *A. aristata*), average	5.5	34.8	39.2	1.0	7.7	17	16	–	–	0.7	35.7	11.7	1.6	1.3	50.8
Arundo donax	8.8	33.0	44.7	1.1	12.4	43	27	48	38	–	–	–	–	–	–
Brachiaria mutica	5.3	34.6	45.7	2.0	12.4	–	–	–	–	–	–	–	2.0	1.7	–
Cenchrus ciliaris	4.9	32.9	51.2	0.8	10.2	35	30	–	–	1.7	51.7	32.4	2.3	1.9	29.3
Chloris barbata,	6.0	30.9	48.8	1.4	11.9	42	61	62	49	2.5	47.5	29.6	2.1	1.7	17.9
pre-flowering															
Chloris gayana	12.3	27.3	45.8	2.4	12.1	60	33	69	54	7.4	55.1	38.6	2.4	2.0	6.5
Young	7.5	37.6	44.3	1.2	10.4	50	33	–	–	3.7	57.8	36.2	2.5	2.1	14.5
Prime	9.4	36.2	42.1	1.2	11.1	55	39	–	–	5.1	58.4	37.0	2.6	2.1	10.4
Ripe	6.4	39.3	44.1	1.3	9.1	49	38	–	–	3.1	53.0	29.8	2.3	1.9	16.1
Chrysopogon fulvus															
(syn. *C. montanus*)															
Pre-flowering	5.9	34.1	48.6	1.6	9.7	34	42	55	50	2.0	46.5	26.8	2.0	1.7	21.9
Post-flowering	4.6	36.6	48.9	1.1	8.6	26	24	–	–	1.2	44.4	2.3	2.0	1.6	37.7
Cynodon dactylon, average	8.4	20.2	56.5	1.4	11.6	–	–	–	–	–	–	–	–	–	–
Cynodon plectostachyus															
Hay	5.4	37.6	45.2	0.9	10.8	–	–	–	–	–	–	–	–	–	–
Pre-flowering	6.4	35.6	50.6	0.6	6.8	–	–	–	–	–	–	–	–	–	–
Dichanthium annulatum															
Pre-flowering	4.6	38.9	46.1	0.9	9.5	28	29	60	42	1.3	44.9	22.3	2.0	1.6	33.5
Flowering	4.1	39.9	44.4	1.0	10.6	–	–	–	–	–	–	–	–	–	–
Seeding	2.7	39.1	45.6	1.2	11.5	–	–	–	–	–	–	–	–	–	–
Hay	9.7	33.0	42.5	1.2	13.6	–	–	–	–	–	–	–	–	–	–

Echinochloa colonum (syn. *Panicum colonum*)	7.3	40.5	43.1	2.0	8.0	–	–	–	–	–	–	–	–	–	–
Eragrostis curvula	–	–	–	–	–	–	–	–	–	6.5	61.0	–	2.6	2.2	–
Erianthus ravennae, flowering	1.9	43.9	45.6	1.6	6.9	0	24	57	44	0.0	45.7	20.2	2.0	1.6	–
Eulaliopsus binata															
Pre-flowering	4.1	38.3	49.3	1.6	7.7	–	–	–	–	–	–	–	–	–	–
Heteropogon contortus															
Flowering	5.6	38.2	46.0	1.4	8.8	–	–	–	–	–	–	–	–	–	–
Young	7.0	34.5	47.9	1.4	9.2	42	42	60	50	2.9	53.0	33.1	2.3	1.9	17.1
Ripe	5.3	36.9	46.9	1.2	9.7	–	36	–	–	0.0	50.7	28.4	2.2	1.8	–
Seeding	3.2	37.4	48.2	1.5	9.7	–	–	–	–	0.0	50.3	25.6	–	–	–
Pre-flowering	4.4	39.6	45.8	1.4	8.8	37	51	–	–	1.6	57.0	34.0	2.5	2.1	33.8
Hymenachne interruppta	7.5	29.3	49.0	1.4	2.9	–	–	–	–	–	–	–	–	–	–
Imperata cylindrica															
Pre-flowering	3.8	39.7	47.9	0.7	7.8	25	26	27	41	0.9	31.9	8.9	1.4	1.1	32.9
Young	4.7	38.1	49.3	1.7	6.3	34	40	59	55	1.6	52.9	39.8	2.3	1.9	32.0
Ipomoea pestigridis	13.6	27.0	44.0	3.5	11.9	57	68	49	57	7.7	49.5	32.3	2.2	1.8	5.4
Ischaemum rugosum	6.6	33.7	47.8	2.2	9.6	42	55	66	49	–	–	26.7	–	–	–
Iseilema laxum															
Pre-flowering	6.4	35.6	50.6	0.6	6.8	53	35	73	61	3.4	57.2	35.0	2.5	2.1	15.3
Lasiurus hirsutus (syn. *L. sindicus*)															
Post-flowering	5.9	38.0	44.7	0.3	11.0	51	0	54	47	3.0	41.4	19.3	1.8	1.5	12.7
Leersia hexandra	6.3	31.4	45.9	1.5	14.9	38	31	66	50	–	–	–	26.6	–	–
Malinis minutiflora															
Pre-flowering	4.4	37.8	48.0	1.0	8.8	37	11	64	62	1.6	55.8	33.7	2.4	2.0	33.7
Flowering	4.2	36.8	48.5	1.1	10.1	31	14	62	58	1.3	52.1	30.1	2.3	1.9	39.4
Paspalum dilatatum	–	–	–	–	–	–	–	–	–	13.9	62.0	–	2.7	2.2	–
Pennisetum clandestinum	–	–	–	–	–	–	–	–	–	6.2	55.0	–	2.4	2.0	–
Pennisetum purpureum	–	–	–	–	–	–	–	–	–	9.2	61.0	–	2.6	2.2	–
Rhynchelytrum repens (syn. *Tricholaena repens*)	5.6	41.6	43.4	1.4	8.0	43	29	63	54	2.4	53.1	28.7	2.3	1.9	21.0

(contd.)

Appendix 3 contd.

Name	Chemical composition on DM basis					Digestibility coefficients				Nutritive value					
	CP	CF	NFE	EE	TA	CP	EE	CF	NFE	kg/100 kg dry matter			Mcal/kg dry matter		NR
										DCP	TDN	NR	DE	ME	
Sehima nervosum	4.5	24.5	–	–	8.9	–	–	–	–	–	–	–	–	–	–
Setaria sphacelata	–	–	–	–	–	–	–	–	–	3.4	58.0	–	2.5	2.1	–
Sporobolus marginatus	6.1	34.4	49.6	1.0	8.8	42	51	–	–	2.6	39.6	19.4	1.7	1.4	14.4
Themeda anathera															
Hay	1.6	39.8	49.6	1.8	7.2	–	–	–	–	–	–	–	–	–	–
Young	16.9	28.5	40.8	1.4	12.7	–	–	–	–	11.2	59.4	42.0	–	–	–
Prime	10.0	35.3	42.6	1.2	10.9	–	–	–	–	5.5	53.1	–	–	–	–
Ripe	6.4	33.3	49.7	0.7	9.9	–	–	–	–	–	–	–	–	–	–
Post-flowering	5.5	35.2	45.7	0.9	12.7	30	54	–	–	3.0	52.8	35.2	–	–	29.1
Themeda arundinacea															
Flowering	1.9	49.0	43.5	1.0	4.6	0	29	57	44	0.0	47.5	19.1	2.1	1.7	–
Urochloa mosambiecensis	–	–	–	–	–	–	–	–	–	5.8	38.5	–	1.7	1.4	–

Source: Sen, et al., 1978

*: Species for which CP, CF, NFE and EE values only are reported are cited from Shankar et al., 1988

**: Species for which DCP, TDN, DE and ME values only are reported are cited form Sharma et al., 1988

DM = Dry matter
CP = Crude protein; CF = Crude fibre; NFE = Nitrogen free extract;
EE = Ether extract; TA = Total Ash; DCP = Digestible crude protein;
TDN = Total digestible nutrients; SE = Starch equivalent; Mcal = Magacalories;
DE = Digestible energy; ME=Metabolizable energy; NR = Nutritive ratio;

APPENDIX 4: Total Ash, Calcium and Phosphorus Content for Tree Leaf Fodders, Shrubs and Grasses

Botanical Name	Per cent on dry matter basis		
	Total ash	CaO	P_2O_5
Tree Leaves and Shrubs			
Acacia catechu	9.80	2.74	0.17
Acica nilotica (syn. *A. arabica*)	5.92-8.22	1.24-1.50	0.11-0.19
Aegle marmelos	14.05	4.24	0.30
Albizia lebbek	7.11-11.54	1.10-2.57	0.15-0.25
Allmania nudiflora	5.46	1.00	0.09
Amorphophallus paeonifolius (syn. *A. campanulatus*	16.22	1.52	0.81
Anogeissus latifolia	9.49	3.03	0.34
Artocarpus heterophyllus, (syn. *A. integrifolia*)	8.33-14.22	0.52-2.15	0.11-0.30
Azadirachta indica	10.15-11.74	0.89-2.76	0.15-0.24
May lopped	7.73	1.39	0.24
September lopped	8.52	2.39	0.19
Bauhinia variegata	9.25-10.09	2.62-2.70	0.25-0.26
Careya arborea	7.48	1.59	0.26
Cordia dichotoma (syn. *C. obliqua*)	14.05	4.24	0.30
Dalbergia sissoo	9.21-13.48	1.95-2.27	0.16-0.21
Dendrocalamus strictus, average	18.33	1.55	0.26
Desmodium oojeinense	10.00	2.40	0.30
Diospyros tomentosa	9.39	1.79	0.16
Ehretia leavis	11.35	1.49	0.27
Ficus bengalensis, average	9.33-14.44	1.81-2.56	0.15-0.28
Ficus racemosa (syn. *F. glomerata*)	11.88-15.14	1.73-2.68	0.16-0.31
Ficus religiosa	13.96-20.91	2.25-4.23	0.21-0.25
Ficus virens (syn. *F. lacor*)			
Young	7.66	1.07	0.25
Mature	13.05	2.21	0.29
Autumn	14.21	2.36	0.14
Fimbristylis quinquangularis	12.51	3.14	0.17
Grewia optiva (syn. *G. oppositifolia*)	–	1.89-3.55	0.16-0.32
Haldina cordifolia (syn. *Adina cordifolia*)	7.61	1.77	0.37
Helicteres isora	10.89	2.25	0.30
Hevea brasilensis	7.29	1.11	0.13
Holoptelea integrifolia	14.52	4.25	0.15
Ipomoea spp.	17.40	1.68	0.46
Kydia calycina	14.41	3.06	0.35
Lagerstroemia parviflora	11.05	2.61	0.25
Lannea coromandelica (syn. *L. grandis*)	9.19	1.75	0.28
Leucaena leucocephala (syn. *L. glauca*)	12.51	3.14	0.17

(contd.)

Appendix 4 contd.

Botanical Name	Percent on dry matter basis		
	Total ash	CaO	P_2O_5
Madhuca latifolia (syn. *M. indica*)	6.72-7.80	1.53-1.58	0.12-0.22
Mallotus philippinensis	8.69	1.31	0.22
Mangifera indica	11.83-13.06	1.87-2.24	0.17-0.33
Millettia extensa (syn. *M. auriculata*)	9.32	1.92	0.29
Mimusops hexandra	7.45	1.40	0.35
Mitragyna parviflora	8.73	2.27	0.24
Morinda pubescens (syn. *M. tinctoria*)	11.48	1.72	0.17
Moringa oleifera	13.43	3.22	0.27
Morus alba	14.32	2.42	0.24
Phoenx dactylifera	8.62	0.60	0.18
Phyllanthus reticulatus	17.94	3.28	0.24
Poinciana elata	13.19	3.55	0.23
Prosopis cineraria (syn. *P. spicigera*)	9.95	2.65	0.24
Quercus floribunda (syn. *Q. dilatata*)	5.11	1.61	0.30
Quercus glauca	7.60	1.87	0.23
Quercus leucotrichophora (syn. *Q. incana*)	5.24	0.99	0.15
Schleichera oleosa	6.15	1.73	0.31
Shorea robusta	3.88	0.77	0.12
Tamarindus indica, average	9.47	2.39	0.25
Monsoon lopped	8.78	1.96	0.23
Winter lopped	13.10	2.37	0.15
Summer lopped	7.76-10.44	1.65-2.08	0.12-0.29
Terminalia bellerica	8.01	2.08	0.27
Terminalia alata (syn. *T. tomentosa*)	10.29	3.20	0.25
Tinospora cordifolia	7.79	1.07	0.57
Zizyphus mauritiana	7.79-11.72	1.42-3.59	0.21-0.33
Zizyphus nummularia	6.20	1.90	0.31
Hays			
Apluda mutica (syn. *A. aristata*)	8.13-14.30	0.34	0.10
Arundo donax	12.40	0.65	0.05
Cenchrus ciliaris	10.18	0.26	0.18
Chloris barbata, Pre-flowering	11.93	0.74	0.15
Chloris gayana			
Young	10.40	0.42	0.20
Prime	11.08	0.30	0.10
Ripe	9.08	–	–
Chrysopogon fulvus (syn. *C. montanus*),	9.73	0.57	0.24
Cynodon dactylon	10.07-14.97	0.35-0.51	0.19-0.26
Cynodon plectostachyus, hay	10.92	0.37	0.11
Pre-flowering	6.78	0.31	0.37

(contd.)

Appendix 4 contd.

Botanical Name	Percent on dry matter basis		
	Total ash	CaO	P_2O_5
Echinochloa colonum (syn. *Panicum colonum*)	7.97	0.39	0.09
Elionesus hirsutus, Post-flowering	11.00	1.09	0.06
Erianthus revennae, Flowering	6.94	0.31	0.06
Eulaliopsis binata, Pre-flowering	7.69	0.36	0.60
Heteropogon contortus			
Flowering	8.76-11.06	0.30-0.34	0.09-0.13
Young	9.22	–	–
Prime	9.22	–	–
Ripe	9.73	–	–
Pre-flowering	13.31	0.34	0.13
Seeding	6.58-10.11	0.28-0.45	0.08-0.10
Hymenachne interrupta	2.90	0.15	0.20
Imperata cylindrica			
Pre-flowering	7.84	0.33	0.10
Young	6.32	0.50	0.15
Ipomoea pestigridis	11.88	1.84	0.34
Ischaemum rugosum	9.63	0.49	0.27
Iseilema laxum, Pre-flowering	6.78-11.22	0.31-0.59	0.10-0.37
Leersia hexandra	14.92	0.14	0.14
Melinis minutiflora			
Pre-flowering	8.83	0.54	0.31
Flowering	10.14	0.37	0.29
Rhynchelytrum repens	7.99	0.34	0.11
Sehima nervosum	8.88	0.51	0.12
Sporobolus marginatus	8.84	0.21	0.14
Themeda anathera	7.20	0.43	0.01
Themeda arundinacea, Flowering	4.58	0.46	0.14

Source: Sen, et al., 1978

APPENDIX 5: Nutritive Value of Tree Leaf Fodders and Grasses as per Detergent System of Analysis

Botanical Name	DM	CP	NDF	CC	ADF	Cellulose	Hemicellulose	Lignin	Silica	Ash
Trees										
Acacia catechu	42.78	12.00	50.60	49.40	35.30	22.80	15.30	11.30	1.60	11.20
Anthocephalus chinensis (syn. *A. cadamba*)	30.00	21.30	34.00	66.00	20.20	11.70	13.80	8.30	0.50	11.08
Azadirachta indica	29.80	13.30	18.52	81.48	13.10	7.00	5.52	5.40	0.86	9.70
Bischofia javanica	48.07	18.00	46.60	53.40	28.40	15.40	18.20	12.40	0.60	15.20
Butea monosperma, (syn. *B. frondosa*)	28.90	17.50	71.40	28.60	45.60	30.30	25.80	11.90	3.20	9.25
Cassia fistula	46.66	15.75	33.20	66.80	23.70	13.00	9.50	10.30	0.40	16.00
Cassia siamea	27.20	13.50	42.80	57.20	25.40	19.40	17.40	4.80	1.20	13.80
Casuarina equisetifolia	43.50	9.45	76.00	24.00	67.00	50.50	9.00	14.30	3.20	21.50
Ficus bengalensis	44.98	11.50	48.80	51.20	42.60	27.80	6.20	9.90	4.90	11.68
Kigelia pinnata	23.50	12.25	56.40	43.60	40.00	26.30	16.40	10.70	3.00	11.00
Madhuca latifolia (syn. *M. indica*)	17.20	10.00	39.20	60.80	28.40	12.24	10.80	4.10	1.77	18.30
Psidium guajava	46.60	9.50	36.00	64.00	33.80	18.20	2.20	13.40	2.20	12.44
Sesbania sesban	28.70	21.00	35.20	64.80	17.60	8.60	17.60	7.60	0.30	10.15
Grasses										
Brachiaria brizantha	36.50	10.50	63.95	36.05	29.20	21.84	34.75	4.23	3.13	–
Brachiaria decumbens	32.07	6.83	61.80	38.20	28.90	22.15	32.90	2.75	4.00	–
Brachiaria mutica	40.74	6.40	64.40	35.60	30.63	23.68	33.77	4.35	2.60	–

Cenchrus ciliaris	45.83	4.73	71.95	28.05	37.43	29.31	34.52	6.30	1.82	–
Cenchrus setigerus	40.00	4.41	71.70	28.30	33.58	23.96	38.12	5.05	4.57	–
Chloris gayana	32.35	7.00	65.20	34.80	32.20	24.67	33.00	3.93	3.60	–
Coix lacryma-jobi	24.26	5.25	71.20	28.80	34.60	23.80	36.60	6.80	4.00	–
Cymbopogon japaus	25.16	6.56	62.30	37.70	32.15	25.45	30.15	4.60	2.10	–
Cymbopogon nardus	22.35	7.00	68.80	31.20	36.00	29.40	32.80	4.60	2.00	–
Cynodon dactylon	31.41	8.14	71.05	28.95	38.13	24.14	32.92	9.75	4.24	–
Dichanthium annulatum	26.97	10.06	68.20	31.80	37.90	31.20	30.30	2.30	4.30	–
Digitaria eriantha	46.00	3.75	67.60	32.40	40.40	30.40	27.20	5.50	4.50	–
Eragrostis chloromelas	49.74	4.38	71.60	28.40	38.80	28.16	32.80	5.72	4.92	–
Eragrostis curvula	26.57	2.69	74.00	26.00	40.00	31.40	34.00	6.80	1.80	–
Eragrostis superba	25.09	3.06	74.00	26.00	41.60	33.00	32.40	6.60	2.00	–
Heteropogon contortus	28.62	2.62	73.80	26.20	37.80	29.00	36.00	4.60	2.40	–
Lasiurus hirsutus (Syn. *L. sindicus*)	35.50	10.19	68.00	32.00	35.40	27.50	32.60	4.40	3.50	–
Panicum antidotale	40.00	7.80	66.60	33.40	35.48	25.93	31.12	5.20	4.35	–
Panicum laevifolium	27.29	10.50	57.05	42.95	28.35	30.25	28.70	4.80	3.30	–
Panicum maximum	25.93	14.62	67.60	32.40	32.50	24.60	35.10	4.60	3.30	–
Paspalum notatum	33.00	7.38	63.90	36.10	29.05	18.55	34.85	5.70	4.80	–
Pennisetum orientale	46.30	7.03	70.10	29.90	33.62	24.84	36.48	5.00	3.78	–
Pennisetum purpureum (Napier grass)	30.10	15.75	57.50	42.50	28.05	19.45	29.45	5.40	3.20	–
Pennisetum purpureum (Hybrid Napier NB.21)	25.48	17.06	61.00	39.00	31.70	23.30	29.30	4.20	4.20	–
Pennisetum purpureum (Napier EB)	28.81	17.50	53.60	46.40	29.25	20.20	24.35	4.85	4.20	–

(contd.)

Appendix 5 contd.

Botanical Name	DM	CP	NDF	CC	ADF	Cellu-lose	Hemi-cellulose	Lignin	Silica	Ash
Pennisetum purpureum (Elephant Grass)	34.00	7.35	71.00	29.00	31.83	26.53	39.17	3.70	1.60	–
Setaria splendida	40.14	5.60	69.00	31.00	32.68	22.10	36.32	3.78	6.80	–
Setaria virdis	26.02	14.00	63.70	36.30	33.40	24.90	30.30	4.40	4.10	–
Tripsacum laxum	32.00	6.12	71.40	28.60	39.60	31.00	31.80	5.00	3.60	–
Urochloa mosambiecensis	48.15	7.43	68.90	31.40	31.35	22.70	37.25	5.30	3.35	–
Urochloa stolonifera	35.16	6.65	65.90	34.10	36.03	27.18	29.85	4.10	4.75	–
Vetiveria zizanioides	25.23	12.68	68.40	31.60	41.25	34.65	27.15	4.40	2.20	–

Source: Pradhan, 1975

DM= Dry matter; CP = Crude protein; NDF =Neutral detergent fibre; ADF = Acid detergent fibre; CC = Cell content

APPENDIX 6: Inventory of Crop and Tree Species in Home Gardens of Kerala State

Crops	Scientific name	Growth form/ economic produce	Average production/ plant or unit area	No. of plants per 0.4 ha
(1)	(2)	(3)	(4)	(5)
Food Crops				
Arrow root	*Maranta arundinacea*	Perennial herb-rhizome	8-10 t/ha	–
Cassava	*Manihot esculenta*	Shrub-tuber	5-15 t/ha	–
Chinese potato	*Coleus parviflorus*	Herb-tuber	10-12 t/ha	–
Dioscorea	*Dioscorea* spp.	Herbaceous climber-tuber	12.5-25 t/ha	18
Rice	*Oryza sativa*	Grass-grain	2 t/ha	–
Sweet potato	*Ipomoea batatus*	Perennial herb-tuber	12-15 t/ha	–
Taro	*Colocasia* spp.	Herb-root	15-20 t/ha	3
Elephant foot yam	*Amorphophallus campanulatus*	Herb-tuber	15-20 t/ha	18
Sugar cane	*Saccharum officinarum*		–	–
Pulses				
Cowpea	*Vigna catjang*	Annual herb-grain	150-350 kg/ha	–
Horse gram	*Dolichos biflorus*	Shrub-grain	600 kg/ha	–
Mung bean	*Vigna aureus*	Annual shrub-grain	300-500 kg/ha	–
Pigeon pea	*Cajanus cajan*	Perennial shrub-grain	450-650 kg/ha	–
Fruits				
Anona	*Annona* spp.	Tree-fruit	50 kg/tree	–
Banana	*Musa* spp.	Perennial herb-fruit	1500 bunches of 20 kg each/ (30 t)/ha	12 –
Bread Fruit	*Artocarpus altilis*	Tree-fruit	300 fruits (1-4 kg each)/ tree/year	–
Garcinia/choisy	*Garcinia indica*	Tree-fruit	100 kg/year	–
Gooseberry	*Emblica officinalis*	Tree-fruit	10-15 kg/tree/year	–
Guava	*Psidium guajava*	Tree-fruit	500-800 fruits/tree/yr	3
Jack fruit	*Artocarpus heterophyllus*	Tree-fruit	500-1000 kg/tree/yr	6
Lemon	*Citrus limon*	Tree-fruit	—	2
Lime	*Citrus aurantifolia*	Tree-fruit	—	1
Amla	*Phyllanthus emblica*	Tree-fruit	—	2
Jamba	*Syzygium jambos*	Tree-fruit	—	1

(contd.)

Appendix 6 contd.

(1)	(2)	(3)	(4)	(5)
Mango	*Mangifera indica*	Tree-fruit	100 kg/tree	12
Papaya	*Carica papaya*	Tree-fruit	30-150 fruits/tree/yr	5
Passion fruit	*Passiflora edulis*	Perennial climber-fruit	20000 fruits/ha	–
Pineapple	*Ananas comosus*	Perennial herb-fruit	—	25
Pomegranate	*Punica granatum*	Perennial shrub-fruit	50 fruits/tree/year	1
Sapota	*Achras zapota*	Tree-fruit	25 kg/tree/year	–
Tamarind	*Tamarindus indica*	Tree-fruit	200 kg/tree/year	2
West India Cherry	*Malpighia punicifolia*	—	—	1
Oil				
Coconut	*Cocos nucifera*	Tree-fruit	40-60 nut/palm/yr	55
Groundnut	*Arachis hypogaea*	Annual herb-kernel	1200-1500 kg pods/ha	–
Sesame	*Sesamum indicum*	Annual herb - seed	250-500 kg/ha	–
Beverages				
Cacao	*Theobroma cacao*	Tree-bean	300 kg dried beans/ha	7
Coffee	*Coffea* spp.	Tree or shrub-bean	800 kg/ha	1
Spices and condiments				
Cardamom	*Elettaria cardamomum*	Perennial herb-seed	100 kg dried capsules/ha	–
Cinnamon	*Cinnamomum zeylanicum*	Tree-bark	150-250 kg quills/ha/yr	–
Clove	*Syzygium aromaticum*	Tree-dried flower buds	2 kg dried buds/tree	–
Ginger	*Zingiber officinale*	Perennial herb-rhizome	10 t/ha (fresh ginger)	500
Nutmeg	*Myristica fragrans*	Tree-fruit	1500-2000 fruits/yr	
Pepper	*Piper nigrum*	Perennial woody climber-seed	2 kg green pepper/vine	23
Turmeric	*Curcuma longa*	Perennial herb-rhizome	3-5 t dry turmeric/ha	–
Curry leaf	*Murraya keonigii*	Tree-leaf		6
Vegetables				
Bhindi	*Abelmoschus esculentus*	Herb-fruit	10 t/ha	–
Bitter gourd	*Momordica charantia*	Climber-fruit	30 t/ha	–
Brinjal	*Solanum melongena*	Herb-fruit	25 t/ha	–
Corneal spinach	*Sauropus androgynus*	—	—	1

Appendix 6 contd.

(1)	(2)	(3)	(4)	(5)
Cucumber	*Cucumis sativus*	Trailing herb-fruit	10-25 t/ha	–
Moringa	*Moringa* spp.	Tree-fruit	10 kg/tree	3
Snake gourd	*Trichosanthes cucumerina*	Climbing annual-fruit	35 t/ha	–
Watermelon	*Citrullus lanatus*	Creeper-fruit	20-35 t/ha	–
Winged bean	*Psophocarpus tetragonolobus*	Perennial climber-fruit	1500 fruits/ha	–
Fodder grasses				
Brazilian lucerne	*Stylosanthes guianensis*	Grass-foliage	10 t/ha	–
Guatemala grass	*Tripsacum laxum*	Grass-foliage	25-30 t/ha	–
Guinea grass	*Panicum maximum*	Grass-foilage	20 t/ha	90
Napier grass	*Pennisetum purpureum*	Grass-foliage	25-35 t/ha	–
Trees				
Bamboo	*Bambusa arundinacea*	Multipurpose	—	–
Erythrina	*Erythrina indica*	Multipurpose	—	–
Gliricidia	*Gliricidia sepium*	Multipurpose	—	–
Leucaena	*Leucaena leucocephala*	Multipurpose	—	–
Mahogany	*Swietenia macrophylla*	Timber	—	–
Morinda	*Morinda tinctoria*	Timber	—	–
Portia tree	*Thespesia populenea*	Timber, fruit	—	–
Teak	*Tectona grandis*	Timber	—	–
Wild jack	*Artocarpus hirsuita*	Multipurpose	—	–
Tree of heaven	*Ailanthus excelsa*	Softwood	—	6
Devils tree	*Alstonia scholaris*	Fuelwood	—	2
Neem	*Azardirachta indica*	Multipurpose	—	2
Mudilla/pezhu	*Barringtonia asiatica*	Stakes	—	7
Silk cotton tree	*Bombax malabaricum*	Fibre, softwood	—	1
India laburnum	*Cassia fistula*	Social, religious, mulch	—	–
Physic nut	*Jatropha curcas*	Stakes	—	4
Boddi/vatta	*Macaranga peltata*	Mulch, stakes, fuelwood	—	4
Pajanelia azantha	*Pajanelia longifolia*	stakes	—	6
Asoka	*Saraca asoca*	Multipurpose	—	1
Strychnine	*Strychnos nuxvomica*	Medicinal	—	1
Others				
Arecanut	*Areca catechu*	Tree-fruit	5kg nuts/palm	3
Betel vine	*Piper betel*	Perennial climber-leaf	2-30 million/ leaves/ha	1

(contd.)

Appendix 6 contd.

(1)	(2)	(3)	(4)	(5)
Cashew	*Anacardium occidentale*	Tree-kernel	5 kg/tree/yr	6
Lemon grass	*Cymbopogon citratus*	Perennial grass-leaf	50-120 kg oil/ha/yr	–
Palmyrah palm	*Borassus flabellifer*	Tree-multipurpose	—	–
Rubber	*Hevea brasiliensis*	Tree-latex	500 kg dried rubber/ha	–
Vetiver	*Vetiveria zizanioides*	Perennial grass-roots	5 t dried roots/ha/yr	–

Source: Nair and Sreedharan, 1986 (without numbers); Sathees Babu et al., 1992 (with numbers)

REFERENCES

Note: To avoid repetition and lengthy citations, acronyms/abbreviations have been used for institutions cited very frequently in these references. The full citations for these abbreviations are as follows:

CAZRI: Central Arid Zone Research Institute, Jodhpur, India

CDS: Centre for Development Studies

CPCRI: Central Plantation Crops Research Institute, Kasargod, India

CSE: Centre for Science and Environment, New Delhi, India

CSWCRTI: Central Soil and Water Conservation Research and Training Institute, Dehradun, India

FRI & C: Forest Research Institute & Colleges, Dehradun, India

ICAR: Indian Council for Agricultural Research, New Delhi, India

ICAR Res. Comp. NEHR: Indian Council for Agricultural Research, Research Complex North Eastern Hill Region, Shillong, India

IGFRI: Indian Grassland & Fodder Research Institute, Jhansi, India

ISTS: Indian Society of Tree Scientists, Solan, India

NICSSR: Northeast India Council for Social Science Research, Shillong, India

Abdul Salam, M., N. Mohan Kumaran, B.K. Jayachandran and three others. 1991. Thirty one tree species support black pepper vine. *Agroforestry Today* 3(4): 16.

Abraham, K.J. 1974. Intercropping in arecanut helps to build up farmers economy. *Arecanut and Spices Bull* 5(3): 73-75 (cited from Muralidharan and Nayar, 1979).

Abraham, P. 1956. Spices as intercrops in coconut and arecanut gardens. *Arecanut J* 7(3): 56-78 (cited from Muralidharan and Nayar, 1979).

Acharya, R.M., B.C. Patnayak and L.D. Ahuja. 1977. Livestock production problems and prospects. In: *Desertification and Its Control,* ICAR, pp. 275-280.

Aggarwal, R.K. 1980. Physicochemical status of soils under *Khejri (Prosopis cineraria* Linn) In: H.S. Mann and S.K. Saxena (eds.), *Khejri (Prosopis cineraria) in the Indian Desert,* CAZRI, pp. 32-37.

Ahuja, L.D. 1980. Grass production under *Khejri* tree. In: ibid., pp. 28-31.

Allan. C.W. 1916. Teak taungya in the Henzada Manbir Division. *Indian Forester* 42 (11 & 12): 533-537.

Alloli, T.B., U.K. Nalawadi and C.V. Patil. 1991. *Casuarina* enhances fruit-tree production on an arid site in India. *Agroforestry Today* 3(4): 12.

Anantha Ram, K. and M.L. Purohit. 1992. The desert agroecosystem of India. In : A. Agarwal (ed.), *Price of Forests,* CSE, pp. 315-319.

Anonymous. 1984. *Research highlights, 1984.* ICAR Res. Comp. NEHR, 41 pp.

Anonymous. 1983. *Shifting Cultivation in North East India.* ICAR Res. Comp. NEHR, 94 pp.

Anonymous, 1982. Editorial. *Agroforestry Systems* 1: 2-7.

Anonymous, 1981. *Proceedings of the Summer Institute on Agroforestry in Semi-arid Zones.* CAZRI.

Anonymous 1979. Package of practices for arecanut. Pamphlet No. 2E, CPCRI.

Anonymous. 1978a. All India Coordinated Coconut and Arecanut Improvement Project. Progress Report for 1976-77, CPCRI, (mimeographed) (cited from Gopalasundaram and Nelliat, 1979).

Anonymous. 1978b. Progress Report for 1977-78, ibid.

Anonymous. 1972. *Indian Forest Utilization, Vol. I & II.* Manager of Publications, Govt. of India, Delhi.

Anonymous. 1964. Intercropping in coconut gardens in Andhra Pradesh. *Coconut Bull* 18: 343-345 (cited from Gopalasundaram and Nelliat, 1979).

Anonymous. 1921 Taungya system as applied to forestry. *Indian Forester* 47 (7): 305-306.

Arora, Y.K. and S.C. Mohan. 1989. Agri-horti systems for wastershed management. *Indian J Soil Conserv* 14 (3): 99-104.

Athreya. 1989. *Pilot Study on Experiences in Farm Forestry in Haryana.* Athreya Management Consultants, New Delhi, India (cited from Saxena, 1991).

Barua, D.N. 1961. Effect of shade and shade trees. *Two and A Bud* 8 (4): 4-12 (cited from Gogoi, 1976).

Barua, D.N. and K.N. Dutta. 1961. Root growth of cultivated tea in the presence of shade trees and nitrogenous manure. *Emp J Exp Agri* 29: 287-298 (cited from Gogoi 1976).

Bavappa, K.V.A. 1980. *Plantation crops. Hand Book of Agriculture,* ICAR, pp. 908-920.

Bavappa, K.V.A., M.K. Nair and T. Prem Kumar (eds.). 1982. *The Arecanut Palm.* Monograph series No. 2, CPCRI, 340 pp.

Bhandary, D.K. 1974. Study of inter and associated crops in areca gardens of the Malnad tract of Karnataka. *Arecanut and Spices Bull* 5(3): 76-77 (cited from Muralidharan and Nayar, 1979).

Bhat, K.S. 1982. Arecanut and cacao. *Annual Report,* CPCRI, pp. 49-50.

Bhat, K.S. 1979. Mixed cropping in arecanut. Tech Bull 3, CPCRI, pp. 35-38 (eds. E.V. Nelliat and K.S. Bhat).

Bhat, K.S. 1974. Intensified inter/mixed cropping in areca garden the need of the day. *Arecanut and Spices Bull 5(3): 67-69* (cited from Muralidharan and Nayar, 1979).

Bhat, K.S. and K.V.A. Bavappa. 1972. Cacao under palms. In: R.L. Wastie and D.A. Earp (eds.), *Cacao and Coconuts in Malaysia,* Proc. Conf. held in Malaysia in 1971, The Incorporated Society of Planters, Kuala Lumpur, Malaysia, pp. 116-121 (cited from Bhat, 1979).

Bhat, K.S. and K.B.A. Khader. 1982. Agronomy (of arecanut). In: K.V.A. Bavappa, M.K. Nair and T. Prem Kumar (eds.), *The Arecanut Palm*, Monograph Series No. 2, CPCRI, pp. 103-131.

Bhat, K.S., K.B.A. Khader, M. Sannamarappa and R.K. Singh. 1983. Inter and mixed cropping trials: arecanut and cacao. *Annual Report*, CPCRI, pp. 59-62.

Bhat, K.S. and M. Leela 1968a. Cultural requirements of arecanut. *Indian Farmg* 18(4): 8-9.

Bhat K.S. and M. Leela. 1968b. Cacao and arecanut are good companions. *Indian Farmg* 18 (4): 19-21.

Bhati, T.K. 1984. Grass component in silvipastoral system with special reference to Indian Arid Zone. In: K.A. Shankarnarayan (ed-in-chief), *Agroforestry in Arid and Semi-arid Zones*, Pub. No. 24, CAZRI, pp. 150-155.

Bhattee, S.S. 1984. Agroforestry in Arunachal Pradesh. In: R.S. Mathur and M.G. Gogate (eds.), *Agroforestry in India*, Proc National Workshop of Agroforestry, FRI & C, pp. 22-26.

Bhimaya, C.P., R.N. Kaul and L.D. Ahuja. 1967. Forage production and utilization in arid and semi-arid rangelands. *Proc on Sci. and India's Food Problem*, National Institute of Science, India, and ICAR, pp. 215-223.

Bhimaya, C.P., R.N. Kaul and B.N. Ganguli. 1964. Studies in lopping intensities of *Prosopis spicigera*. *Indian Forester* 90 (1): 20-23.

Bjorkman, E. 1942. *Symbolae botanicae. Upsaliensis* 6:1-190 (cited from Biol Absts 20: 7878(1946) by Wight, 1959).

Bohra, H.C. and P.K. Ghosh. 1980. The nutritive value and digestibility of *loong*. In: H.S. Mann and S.K. Saxena (eds.), *Khejri (Prosopis cineraria) in the Indian Desert*, CAZRI, pp. 45-50.

Boutland, A., M. Robinson, J. Field, M. Scheltema, B. Hawkins, P. Bulman, J. Scarvelis, B. Mason, M. Wallace, M. Castley, P.A. Ryan, R.T. Henderson, G.B. Applegate, C. Lansdown and R. Reid. 1992. Alternative products from trees and shrubs to "The Role of Trees in Sustainable Agriculture" Conference. *Agroforestry Systems* 20: 25-58.

Brahma, R.N. 1973. In north Bengal banana is paying intercrop in areca gardens. *Arecanut and Spices Bull* 5 (3): 80-81 (cited from Muralidharan and Nayar, 1979).

Brandis, D. 1906. *Indian Trees* (Reprinted 1971). Bishan Singh Mahendra Pal Singh, Dehradun, India, 767 pp.

Branford, D. 1922. *Agric. J India* 17: 242 (cited from Sen, 1953)

Buckingham, J. 1885. Papers regarding the *sau* tree and its remarkable influence on tea bush. Indian Tea Assoc, Calcutta, India (cited from Wight, 1959).

Caleda, A. and I.D. Esteban. 1981. Agroforestry in the Philippines. Paper presented at Workshop on Environmentally Sustainable Agroforestry and Fuelwood Production with Fast Growing, Nitrogen Fixing, Multipurpose Legumes, Env. & Policy Inst., East West Center, Honolulu, U.S.A. (mimeographed), 34 pp.

Cardamom Board. 1986. *Cardamom Culivation*. Cardamom Board, Cochin, India, 38 pp.

CDS. 1990. A study on tribal economy and its dependence on minor forest produce and agroforestry system, in six tribal districts of Orissa. CDS, Bhubaneshwar, India (mimeographed).

Census of India. 1982. Primary Census Abstract. General Population Series 1, part II B (I), Census Commission of India, Controller of Publications, Delhi, India.

Champion, H.G. and A.L. Griffiths. 1948. *Mannual for General Silviculture for India*. Revised edition. Oxford University Press, Delhi, India, 330 pp.

Chatterjee, B.N., S. Maiti, S.K. Majhi and S. Pal. 1988. Aerial biomass production of *Anjan (Cenchrus ciliaris* L) grass in *Subabul* (*Leucaenia leucocephala* (Lam de Wit) alleys in subhumid Gangetic plains. In: Punjab Singh and R.S. Pathak (eds.), *Rangelands Resources and Management*, Range Management Soc., India, IGFRI, pp. 356-362.

Chaturvedi, A.N. 1992. Optimum rotation of harvest of poplars in farmlands under agroforestry. *Indian Forester* 118: 81-82.

Chaturvedi, M.D. 1948. The role of leaf fodder in the United Provinces. U.P. Forest Dept, Bull No.19, 9 pp.

Chempakam, B., S.T.K. Annamalai and K.N. Murthy. 1982. Other uses of arecanut. *Indian Farmg* 32(9): 40-43.

Chinnamani, S. 1983. Personal communication.

Chinnamani, S., B.L. Rathore and M.C. Prajapati. 1982. Scientific management and production potential of tree and grass species in lare scale plantations on sub-catchment basis in Chambal ravines at Kota. Annual Report, CSWCRTI, pp. 84-87.

Cincotta, R. and G. Pangare. 1993. Pastoralists: Brokers of agricultural soil fertility. *Wastelands News* 8(3):40-44.

Clements, H.F. 1928. *Plant Physiol* 3:441-458 (cited from Wight, 1959).

Clousten, D. and F.J. Plymen. 1920. *Agric J India* 15: 190 (cited from Sen, 1953).

Coffey, T.M. 1929. Hill taungyas in Kurseong division, North Bengal. *Indian Forester* 55 (2): 80-81.

Combe, J. 1982. Agroforestry techniques in tropical countries: potential and limitations. *Agroforestry Systems* 1(1): 13-28.

Cooper, H.R. 1939. *Nitrogen Supply to Tea.* Indian Tea Assoc, Calcutta, India (cited from Wight, 1959a).

CPCRI. 1986a. Arecanut: package of practices. Pamphlet No. 2, CPCRI, 21 pp.

CPCRI. 1986b. Cacao: package of practices. Pamphlet No. 4, ibid., 18 pp.

CPCRI. 1986c. Pepper: packages of practices. Pamphlet No. 5, ibid., 14 pp.

CPCRI. 1984. Multistoried cropping at Kasargod. Annual Report, ibid., pp. 28-30.

CPCRI. 1982. Ibid., pp. 28-30.

CPCRI. 1981. Annual Report for 1980, ibid.

CPCRI. 1979. Annual Report for 1978, ibid.

CPCRI. 1977. Annual Report for 1976, ibid.

CPCRI. 1974. Annual Report for 1973, ibid., pp. 27.

CPCRI. 1969. Annual Report for 1968, ibid. (mimeographed).

CSWCRTI. 1981. Report on Operational Research Project: Watershed Management. CSWCRTI, 144 pp.

DAM. 1943. Madras Agricultural Stations Report for 1941-42, Department of Agriculture, Madras, India, pp. 323-334 (cited from Gopalsundaram and Nelliat, 1979).

DAM. 1942. Madras Agricultural Stations Report for 1940-41, ibid., pp. 375-384.

DAM. 1941. Madras Agricultural Stations Report for 1939-40, ibid., pp. 367-376.

DAM. 1934. Madras Agricultural Stations Report for 1932-33, ibid., pp. 267-276.

DAM. 1932. Madras Agricultural Stations Report for 1930-31, ibid., pp. 248-266.

Das, G.M. 1961. Studies on the effect of shade and manure on the incidence of red spider. Annual Report, 1960, Tocklai Experiment Station, Jorhat, India, pp. 111-112.

Das, M.C. and S. Mahapatra. 1984. Agroforestry practices in Orissa. In: R.S. Mathur and M.G. Gogate (eds.), *Agroforestry in India,* op. cit., FRI & C, pp. 154-156.

Das, P.K. 1984. Estimating production costs and returns for coconut in Kerala. *J Plant Crops* 2 (2): 52-59.

Das, P.K. 1982. Economics of multistoried cropping in coconut. Annual Report, CPCRI, pp. 87.

Dayal, R. 1974. Report on achievements of Soil Conservation Reseach, Demonstration and Training Centre, Vasad, India (mimeographed).

DCD. 1970. Coconut bulletin. Directorate of Coconut Development, Govt. of India, Cochin, India (cited from Nair, 1979).

De, R.N. 1932. Taungya in Garo hills division, Assam. *Indian Forester* 58 (2): 93-99.

Devashayam, R.S. and C.P.R. Nair. 1982. Insect pest management in arecanut. *Indian Farmg* 32 (9): 44-46.

Dhanda, R.S. 1984. Agroforestry in Punjab. In: R.S. Mathur and M.G. Gogate (eds.), *Agroforestry in India,* op. cit., FRI & C, pp. 178-185.

Dhillon, G.S., S.S. Grewal and A.S. Atwal. 1979. Developing agrisilviculture practices: Effect of farm trees (*Ecalyptus*) on the adjoining crops. *Indian J Ecol* 6: 88-97.

Dhillon M.S., Surjeet Singh., A.S. Atwal and G.S. Dhillon. 1984. Agrisilvicultural practices. Effect of *Dalbergia sissoo* and *Acacia nilotica* on the yield of adjoining crops. *Indian J Ecol* 11(2): 249-253.

Dhillon, G.S., Surjeet Singh., M.S. Dhillon and A.S. Atwal. 1982. Developing agrisilviculture practices: Studies on shading effect of *Ecalyptus* on the yield of adjoining crops. *Indian J Ecol* 9(2): 228-236.

Dhir, R.P. 1977. Western Rajasthan soils: their characteristics and properties. In: *Desertification and Its Control,* ICAR, pp. 112-115.

Dogra, A.S. 1984. Farm and agroforestry in southwest Punjab. In: R.S. Mathur and M.G. Gogate (eds.), *Agroforestry in India,* op. cit., FRI & C, pp. 157-177.

Drummond, R.O. 1944. Lopping in the Kumaon Himalaya of United Provinces. *Indian Forester* 70(12): 401-418.

Dutta, A.C. 1977. Shade trees, green crops and cover crop plants in the tea estates of North East India. Memorandum No. 30, Toklai Experiment Station, Jorhat, India, 128 pp.

Dutta, B.K. and R.P. Dhir. 1981. Mineral elements in relation to animal requirements. In: H.S. Mann and S.K. Saxena (eds.), Bordi (*Zizyphus nummularia*), CAZRI, pp. 34-37.

Eernstman, T. 1968. Chemical analysis of leaves and other organs of *Theobroma cacao* as a means of diagnosing fertilizer requirements. Pub No 54, PUDOC, Wageningen, Netherlands, 55 pp.

Fremond, Y., R. Ziller and M.N. Lamothe. 1966. *The Coconut Palm.* Potash Institute, Berne, Switzerland.

FRI & C. (year not known). *Peoples Programme for Tree Planting.* FRI & C.

Gangolly, S.R., R. Singh, S.L. Katyal and D. Singh. 1957. *The Mango.* ICAR, 530 pp.

Ganguli, B.N., R.N. Kaul and K.T.N. Nambiar. 1964. Preliminary studies on a few top feed species. *Annals Arid Zone* 3 (1&2): 33-37.

Ganguly, J.B. 1980. Socio-economic problems of transition from shifting cultivation to sedentary cultivation in northeast India. In: *Shifting Cultivation in Northeast India,* NICSSR, pp. 28-36.

Ghildyal, B.N. 1991. Oak-Production, Management and Use in the Himalayas. Field Doc No. 27, Regional Wood Energy Dev. Prog. Asia, FAO Regional Office, Bangkok, Thailand, 46 pp.

Ghildyal, B.P. 1981. Soils of the Garhwal and Kumaon Himalayas. In: J.S. Lal (ed.), The Himalayas: *Aspects of Change,* Oxford University Press, Delhi, India, pp. 120-151.

Ghose, B., S. Singh and A. Kar. 1977. Geomorphology of Rajasthan desert. In: *Desertification and Its Control,* ICAR, pp. 102-115.

Glausson, A.K. 1927. Regeneration work in northern Bengal. *Indian Forester* 53(2): 71-85.

GOG. 1984. Gujarat wood balance study. Government of Gujarat, Ahmedabad, India (cited from Saxena, 1991).

Gogoi, B.N. 1976. A review of research in shade tea in India. *Two and A Bud* 23 (2): 67-73.

Goodwin-Austen, H.H. 1875. A celt found at the Khasi hills at Shillong. Proc Asiatic Soc. Bengal (cited from Sharma, 1980).

Goor, A.Y. and C.W. Barney. 1976. Arid Zones. In: *Forest Tree Planting in Arid Zones.* Second edition, Ronald Press, New York, USA, pp. 3-53 (cited fom Alloli et al., 1991).

Gopalasundaram, P. and E.V. Nelliat. 1979. Intercropping in coconut. In: E.V. Nelliat and K.S. Bhat (eds.), Multiple cropping in coconut and arecanut gardens, Tech Bull 3, CPCRI, pp. 6-23.

Gorrie, R.M. 1937. Tree lopping on a permanent basis. *Indian Forester* 63(1): 29-37.

Grewal, S.S. 1992. Watershed management project, Relmajra (Punjab). Annual Report, 1991-1992, CSWCRTI, pp. 142-143.

Grewal, S.S., S.P. Mittal, S. Dayal and Y. Agnihotri, 1992. Agroforestry systems for soil and water conservation and sustainable production from foothill areas of north India. *Agroforestry systems* 17: 183-191.

Griffith, A.L. 1942. Teak plantation technique. *Indian Forest Rec(ns) Silvi* 5(2) (cited from Sangal, 1984.)

Grimwood, 1975. *Coconut Palm Products.* Food and Agriculture Organization, Rome, Italy.

Gunjal, S.S. and P.L. Patil. 1992. Mycorrhizal control of wilt in *casuarina*. *Agroforestry Today* 4(2): 14-15.

Gupta, B.L. 1956. *Forest Flora of Chakrata, Dehradun and Saharanpur Forest Divisions.* Third edition. Manager of Publications, Govt. of India, Delhi, India, 558 pp.

Gupta, B.N. 1991. Status of non-wood forest products in India. Regional Expert Consultation on Non-Wood Forest Products in the Asia-Pacific Region, FAO Regional Office for Asia and the Pacific, Bangkok, Thailand (mimeographed), 63 pp.

Gupta, B.N. 1981. *Dairy Cattle Feed Resources and Feeding Practices in the North Eastern States.* Pub No. 191, National Dairy Res. Inst., Karnal, India, 161 pp.

Gupta, G.P. 1975. Sediment production status report on data collection and utilization. *Soil Conserv Digest* 3(2): 10-20.

Gupta, P.N. 1983. Cardamom in Sikkim State Personal communication.

Gupta, P.N. 1982. Export potential in large cardamom. *Indian Hort* 26(4): 21-25.

Gupta, R. 1984. Integration of medicinal plants cultivation in forest and forest plantations of north western Himalaya. In: P.K. Khosla, Sunil Puri and D.K. Khurana (eds.), *Agroforestry Systems—A New Challenge,* Proc. Satelite Seminar on Agroforestry Systems—A New Challenge, ISTS.

Gupta, T. 1979. Some financial and natural resource management aspects of commercial cultivation of irrigated *Eucalyptus* in Gujarat, India. *Ind J For* 2(2): 118-137.

Gupta, T. and A. Guleria. 1982. *Non-wood Forest Products in India.* Oxford & IBH Publishing Co. Pvt. Ltd., New Delhi, India, 147 pp.

Hadfield, W. (year not known). New Thoughts on shade. A leaflet by Indian Tea Association, Calcutta, India.

Hadfield, W. 1974. Shade in northeast Indian tea plantations. I. Shade pattern. *J Appl Ecol* 11: 151-178.

HAPPRC. Promoting broadleaf tree species for mountains. Information Bull 1, High Altitude Plant Physiology Res. Centre, Garhwal University, Srinagar, India.

Hay, E.F.A. 1924. Sugar cane in taungya. *Indian Forester* 50 (12): 644-645.

Hazra, C.R. 1985. Forage and seed yields of crops under agroforestry production system—radiation and temperature relations. *J Agron & Crop Sci* 155: 186-192.

Hazra, C.R. and S.B. Tripathi. 1986. Soil properties, micrometeorological paramaters, forage yield and phosphorus uptake of berseem as influenced by phosphate application under agroforestry system of production. *J Agron & Crop Sci* 156: 145-152.

Heath, E.D. 1956. Drought and tea. Indian Tea Assoc. Sci. Dept, Cachar Advisory Branch Memo, 13pp (cited from Wight, 1959).

Hooda, R.S. 1984. Agroforestry and rural development in Haryana State. In: R.S. Mathur and M.G Gogate (eds.), *Agroforestry in India,* op. cit., FRI & C, pp. 55-64.

Howard, S.H. 1939. Effect of standards on field crops in taungya. *Indian Forester* 65(3): 151-152.

IAB. 1947. *The Use and Misuse of Shrubs and Trees as Fodder.* Joint Pub. No. 10, Imperial Bureau Pastures and Field Crops, Imperial Agricultural Bureau, Aberswyth, England.

ICRI. 1987. Indian Cardamom Res. Inst. Myladumpara, India. Personal communication.

IMD. 1982. Climate of north east India. In: K. Nandi (ed.), *Resource Potentials of North East India,* Indian Meteorological Dept., Meghalaya Sci. Soc., Shillong, India, pp. 15-20.

Jambulingam, R. and E.C.M. Fernandes. 1986. Multipurpose trees and shrubs of farmlands in Tamil Nadu State, India. *Agroforestry Systems* 4: 17-32.

Jeevarathnam, K., P. Samraj, and C. Henry. 1992. Arresting root invasion. *The Hindu,* Nov. 25, 1992, Madras, India.

Jodha, N.S. 1977. Land tenure problems and policies in the arid region of Rajasthan. In: *Desertification and Its Control,* ICAR, pp. 335- 347.

Kailasam, C., P. Gopalasundaram and E.V. Nelliat. 1983. Mixed cropping with cacao. Annual Report, CPCRI, pp. 58-59.

Kannan, R. and K.P.P. Nambiar. 1976. Studies on intercropping coconut garden with annual crops. *Coconut Bull* 5: 1-3 (cited from Gopalasundaram and Nelliat, 1979).

KAU. 1979. Research report for 1977-78. Kerala Agricultural University, Trichur, India (cited from Gopalasundaram and Nelliat, 1979).

Kaul, R.N. and B.N. Ganguli. 1963. Fodder potential of *Zizyphus* in the scrub grazing lands of arid zone. *Indian Forester* 89: 623-630.

Kaul, R.N. and R.P. Kapoor. 1992. Unpublished data. Personal communication.

KGSN. 1985. Balanced (small) farm—an appropriate garden technology. Phamphlet, Kerala Gandhi Smarak Nidhi, Trivandrum, India (cited from Nair and Sreedharan, 1986).

Khader, K.B.A. 1982. Inter and mixed cropping in areca gardens. *Indian Farmg* 32 (9): 21-23.

Khader, K.B.A. and K.J. Antony. 1968. Intercropping: a paying proposition for areca growers-what crops to grow. *Indian Farmg* 8 (4): 4-5 (cited from Muralidharan and Nayar, 1979).

Khan, M.I.R. 1955. Tropical thorn forest of West Pakistan. *Pakistan J Forestry* 5: 161-171.

Khosla, P.K. 1984. Agroforestry resources of western Himalayas. In: R.S. Mathur and M.G. Gogate (eds.), *Agroforestry in India,* op. cit., FRI & C, pp. 274-302.

Khybri, M.L., R.K. Gupta and Sewa Ram. 1983. Effect of *Grewia optiva, Morus alba* and *Eucalyptus hybrid* on the yield of crops under rainfed conditions. Annual Report, CSWCRTI, pp. 61-62.

Khybri, M.L., R.K. Gupta, Sewa Ram and H.S. Tomar. 1992. Crop yields of rice and wheat grown in rotation as intercops with three tree species in the outer hills of Himalaya. *Agroforestry Systems* 19: 193-204.

Khybri, M.L. and Sewa Ram. 1984. Effect of *Grewia optiva, Morus alba* and *Eucalyptus hybrid* on the yield of crops under rainfed conditions. Annual Report, CSWCRTI, pp. 51-53.

Khybri, M.L., Sewa Ram and S.P. Bhardwaj. 1988. Study on intercropping of field crops with fodder crop of *subabul* (*Leucaena leucocephala*) under rainfed conditions. In: Ibid., pp. 53-54.

Koul, K.K., A. Parida, M.S. Bisht and S.N. Raina. 1991. Variability in *Prosopis cineraria* from India's Thar desert. *Agroforestry Today* 3(4): 12.

Krishna Marar, M.M. 1964. Intercropping in coconut plantations in India—problems and prospects. FAO Tech Wkg Party Coconut Production, Colombo, Sri Lanka, pp. 393-398 (cited from Gopalasundaram and Nelliat, 1979).

Krishnaji, N., Chandhan Mukerji, E.V. Nelliat, P.K.R. Nair, Jacob Mathew and P.T. Varghese. 1976. Economics of inter and mixed cropping in the coconut gardens of Kerala : some preliminary findings. CDS, Trivandrum, India, Working Paper No. 41 (mimeographed) (cited from Gopalasundaram and Nelliat, 1979).

Krishnan, A. 1977. A climatic analysis of the arid zone of northwest India. In: *Desertification and Its Control,* ICAR, pp. 42-57.

KSPB, 1973. An evaluation report on oil palm station, Thodupuzha. Evaluation Series No. 18. Kerala State Planning Board, India, (cited from Nair, 1979).

Kumar, L.S.S. 1946. Better utilization of forest grasslands in Bombay Province. *Indian Forester* 72: 162-164.

Kushwah, B.L., E.V. Nelliat, V.T. Markose and A.F. Sunny. 1973. Rooting pattern of coconut (*Cocos nucifera* L). *Indian J Agron* 18: 71-74.

Kutty, M.G. 1980. Taungya cultivation in Kerala. Second Forestry Conference. FRI & C (cited from Sangal, 1984).

Lahiri, A.K. 1984a. Agroforestry in West Bengal (Part I). In: R.S. Mathur and M.G. Gogate (eds.). *Agroforestry in India,* op. cit., FRI & C, pp. 218-224.

Lahiri A.K. 1984b. Agroforestry in West Bengal (Part II). In: ibid., pp. 225-231.

Lahiri, A.N. 1980. *Prosopis cineraria* in relation to soil and other conditions of its habitat. In: H.S. Mann and S.K. Saxena (eds.), *Khejri* (*Prosopis cineraria*) *in the Indian Desert,* op. cit., CAZRI, pp. 33-44.

Leela, K. and U.P. Bhaskaran. 1978. Effect of intercropping coconut stands with groundnut on soil fertility and plantation management. *Proc Placrosym* I, Placrosym Standing Committee, Kasargod, India, pp. 393-398.

Lundgren, B. 1982. Cited from *Agroforestry Systems* 1:2-7.

Majumdar, D.N. 1980. Shifting cultivation, is it a way of life? An analysis of Garo data. In: *Shifting Cultivation in north east India,* NICSSR, pp. 14-18.

Maikhuri, R.K. and P.S. Ramakrishnan. 1990. Ecological analysis of a cluster of villages emphasizing land use of different tribes in Meghalaya in northeast India (cited from Ramakrishnan, 1992).

Malhotra, S.P. 1984. Traditional agroforestry practices in arid zone of Rajasthan. In: K.A. Shankaranayan (ed-in-chief), *Agroforestry in Arid and Semi-arid Lands*, Pub. No. 24, CAZRI, pp. 263-266.

Malhotra, S.P. 1977. Socio-demographic factors and nomadism in the arid zone. In: *Desertification and Its Control*, ICAR, pp. 310-323.

Malwade, P.A. 1984. Farm forestry in Gujarat. A successful programme. In: R.S. Mathur and M.G. Gogate (eds.), *Agroforestry in India*, op. cit., FRI & C, pp. 48-54.

Mann, H.S., S.P. Malhotra and K.A. Shankarnarayan. 1977. Land and resource utilization in the arid zone. In: *Desertification and Its Control*, ICAR, pp. 89-101.

Mann, H.S. and S.K. Saxena. 1980. *Khejri* (*Prosopis cineraria*) *in the Indian Desert*, Monograph No.11, CAZRI, 78 pp.

Mann, H.S. and R.P. Singh. 1977 Crop production in the Indian arid zone. In: *Desertification and Its Control*, ICAR, pp. 215-224.

Manning, D.E.B. 1941. Some aspects of the problems of Taungya in Burma. *Indian Forester* 67(10): 502-505.

Masters, J.W. 1863. *J Agric Hort Soc, India* 13: 31-47 (cited from Wight, 1959a).

Mathew, J. and M. Shaffee. 1979. Mixed farming in coconut gardens. In: E.V. Nelliat and K.S. Bhat (eds.), Tech Bull 3, CPCRI, pp. 42-46.

Mathew, P.M. 1985. "Chemmeen" cultivation in "Pokkali" lands. Kerala Agricultural University, Trichur, India, *Kalpadhenu* 13 (1) (cited from Nair and Sreedharan, 1986).

Mathew, T., B. Mohan Kumar, K.V. Suresh Babu and K. Uma-maheshwaran. 1992. Comparative performance of four multipurpose trees associated with four grass species in the humid regions of Southern India, *Agroforestry Systems* 17: 205-218.

Mathur, C.M., S.N. Ganu, V.B. Moghe and S.V. Jain. 1972. *Soils of Rajasthan*. Soil Survey Organisation, Dept. of Agriculture, Jaipur, India.

Mathur, H.N. and P. Joshie. 1972. Annual Report, CSWCRTI, (mimeographed).

McConnel, D.G. and K.A.B. Dharampala. 1973. The economic structures of Kandyan forest farms. Farm Management Report 7. UNDP/FAO Agricultural Diversification Project, Peredenya, Sri Lanka.

Menon, K.P.V. and K.M. Pandalai. 1960. *The Coconut Palm—A Monograph*. Indian Central Coconut Committee, Ernakulam, India.

Mia, W.H., B. Sahai, B.N. Majumdar and N.D. Kehar. 1960. Studies on tree leaves as cattle fodder. Part III. The nutritive value of *bargad* leaves (*Ficus bengalensis*). *Indian J Dairy Sci* 13(1): 1-8.

Mishra, B.K. and P.S. Ramakrishnan. 1983a. Slash and burn agriculture at higher elvations in north eastern India. I. Sediment, water and nutrient losses. *Agric Ecosyst and Env* 9: 69-82.

Mishra, B.K. and P.S. Ramakrishnan. 1983b. Slash and burn agriculture at higher elevations in north eastern India. II. Soil fertility changes. *Agric Ecosyst and Env* 9: 83-96.

Mishra, B.K. and P.S. Ramakrishnan. 1982. Energy flow through a village ecosystem with slash and burn agriculture in north east India. *Agric Syst* 9: 57-72.

Mishra, B.K. and P.S. Ramakrishnan. 1981. The economic yield and energy efficiency of hill agro-ecosystems at higher elevations of Meghalaya in north eastern India. *Acta Oecologica Oecol Applic* 2(4): 369-389.

Mishra, J. and U.N. Prasad. 1980. Agrisilviculture studies on raising of oilseeds like *til*, groundnut and soybeans as cash crops in conjunction with *Dalbergia sissoo, Tectonia grandis* at Mandar, Ranchi. *Indian Forester* 106(12): 829-836.

Mishra, P.R. and A.D. Sud. 1984. Tie ridge technique of raising *Eucalyptus* and *Bhabar* on 4% slope. Annual Report, CSWCRTI, pp. 84-85.

Mishra, P.R. and A.D. Sud. 1983. Tie ridge technique of raising *Eucalyptus* and *Bhabar* on 4% slope. Annual Report, CSWCRTI, pp. 57-60.

MOA. 1992a. *Horticultural Statistics*. National Hort. Board, Ministry of Agriculture, New Delhi, India.

MOA. 1992b. *Indian Agriculture in Brief*. Ministry of Agriculture, New Delhi, India

MOA. 1985. Ibid.

MOA. 1983. Statewise area and production of important fruits in India from 1976-77 to 1978-79 and projections up to 1982-83. Ministry of Agriculture, New Delhi (mimeographed).

MOA. 1982. *Indian Agriculture in Brief.* Op. cit.

Momin, S.A. and S.C. Ray. 1943. Tree leaves as cattle fodder I. The seasonal variation in the composition of some edible leaves. *Indian J Vet Sci* 13: 183-190.

Muralidharan, A. and T.V.R. Nayar. 1979. Intercropping in arecanut gardens. In: E.V. Nelliat and K.S. Bhat (eds.), Tech Bull 3, CPCRI, pp. 24-27.

Murthy, K.N. and R.S.N. Pillai. 1982. Botany (of arecanut). In: K.V.A. Bavappa., M.K. Nair, and Prem Kumar (eds.), Monograph Series No. 2, CPCRI, pp. 1-49.

Muthana, K.D., 1980. Silviculture aspects of *Khejri.* In: H.S. Mann and S.K. Saxena (eds.), *Khejri (Prosopis cineraria) in the Indian Desert,* CAZRI, pp. 20-24.

Muthana, K.D., S.K. Sharma, A.K. Raina and G.L. Meena. 1980. Silvipastoral studies. Annual Report, 1980, CAZRI (cited from Bhati, 1984).

Nagaraj, B. 1974. Intercropping in arecanut gardens of Maidan areas of Karnataka. *Arecanut and Spices Bull* 5(3): 78-79 (cited from Muralidharan and Nayar, 1979).

Naidu, G.V.B. 1959. Betelvine cultivation in areca gardens and its benefits. *Arecanut J* 9(4): 128-130 (cited from Muralidharan and Nayar, 1979).

Nair, C.T.S. and C.N. Krishnankutti. 1984. Socio-economic factors influencing farm forestry. A case study of tree cropping in homesteads in Kerala, India. In: Y.S. Rao, N.T. Vergara and G.W. Lovelace (eds.), *Community Forestry: Socio-economic Aspects,* FAO Regional Office for the Asia-Pacific Region, Bangkok, Thailand, pp. 115-130.

Nair, K.K. 1981. Agroforestry in the southern hills. In: *Proc Agroforestry Seminar,* ICAR, pp. 234-241.

Nair, M.A. and C. Sreedharan. 1986. Agroforestry farming systems in the homesteads of Kerala, southern India. *Agroforestry Systems* 4: 339-363.

Nair, M.G. 1982. Intercropping with pepper. *Indian Farmg* 32(9): 17-19.

Nair, P.K.R. 1985. Classification of agroforestry systems. *Agroforestry Systems* 3: 97-128.

Nair, P.K.R. 1979. *Intensive Multiple Cropping with Coconuts in India.* Verlaug Paul Parey, Berlin and Hamburg, Germany, 147 pp.

Nair, P.K.R. 1976. Intensive cropping for stabilized income at enhanced rates from coconut plantations. *Financing Agri* 7(3): 3-8.

Nair, P.K.R., and T.K. Balakrishnan. 1976. Pattern of light interception by canopies in a coconut cacao crop combination. *Indian J Agric Sci* 46: 453-462.

Nair, P.K.R., E.C.M. Fernandez and P.N. Wambugu. 1984. Multipurpose leguminous trees and shrubs for agroforestry. *Agroforestry Systems* 2: 145-163.

Nair, P.K.R., Rama Varma and E.V. Nelliat. 1974. Intercropping for enhanced profits from coconut plantation. *Indian Farmg* 24 (4): 11-13.

Nair, R.V., N.G. Pallai and K.N. Sahasranaman, 1976. Mixed farming in coconut gardens. *Indian Farmg* 26 (11): 24-25.

Nambiar, K.K. 1949. A Survey of Arecanut Crop in Indian Union. Indian Central Arecanut Committee, Calicut, India.

Nambiar, K.T.N, 1992. Effect of closure on ecological succession and natural regeneration in ravine lands. Annual Report, 1991-92, CSWCRSTI, pp. 105-106.

Nambiar, K.T.N. and H.B. Singh. 1991. To study the effect of raising fast growing tree species around agricultural crops on crop production. Annual Report, 1990–91, ibid., pp. 96-98.

Nambiar, K.T.N. and H.B. Singh. 1990. To study the effect of raising fast growing tree species around agricultural crops on crop production. Annual Report, 1989-90, ibid., pp. 124.

Nambiar, K.T.N., H.B. Singh and S. Chinnamani. 1988. To study the effect of raising fast growing tree species around agricultural crops on crop production. Annual Report, 1988, ibid., pp. 159-160.

Nambiar, K.T.N., H.B. Singh and S. Chinnamani. 1987. To study the effect of raising fast growing tree species around agricultural crops on crop production. Annual Report, 1987, ibid., pp. 148-149.

Nambiar, K.T.N., H.B. Singh and S. Chinnamani. 1986. To study the effect of raising fast growing tree species around agricultural crops on crop production. Annual Report, 1986, ibid., pp. 146-147.

Nayar, T.V.R. 1982. Grow pepper in arecanut gardens. *Indian Farmg* 32 (1): 27-31.

NCA. 1976. Report of the National Commission on Agriculture. National Commission of Agriculture, Govt. of India, New Delhi, India, Part IX, pp. 427-472.

NEC. 1982. *Shifting Cultivation in North Eastern Region.* Pub. No. 17, North Eastern Council, Shillong, India, 24 pp.

Nelliat, E.V. 1979a. Prospects of multiple cropping. In: E.V. Nelliat and K.S. Bhat (eds.), Tech Bull 3, CPCRI, pp. 3-5.

Nelliat, E.V. 1979b. Multistoried cropping. In: ibid., pp. 39-41.

Nelliat, E.V., K.V.A. Bavappa and P.K.R. Nair. 1974. Multistoried cropping. A new dimension in multiple cropping for coconut plantations. *World Crops* 26: 262-266.

Nelliat, E.V. and K.S. Bhat. 1979. Multiple cropping in coconut and arecanut gardens. Tech Bull 3, CPCRI, 54 pp.

Nelliat, E.V., P. Gopalasundaram, P.T. Varghese and K.S. Sivaraman. 1979. Mixed cropping in coconut. Ibid., pp. 28-35.

Nelliat, E.V. ad N. Krishnaji. 1976. Intensive cropping in coconut gardens. *Indian Farmg* 27(9): 9-12.

Ninan, K.N. 1992. Economics of shifting cultivation. In : A. Agarwal (ed.), *The Price of Forests,* CSE, pp. 311-314.

Pal, R.N., K.K. Dogra, L.N. Singh and S.S. Negi. 1979. Chemical composition of some fodder trees in Himachal Pradesh. *Forage Res* 5: 109-115.

Patil, B.D., R. Deb Roy and P.S. Pathak. 1981. Agroforestry research and development with reference to Indo-Gangetic Plains. In: *Proc Agroforestry Seminar,* ICAR, pp. 40-68.

PC. 1985. *Report of the Working Group on Hill Areas.* Planning Commission, Govt. of India, New Delhi, India (mimeographed), 446 pp.

PC. 1982. *Socio-economic Development of Himalayan Hills.* Task Force for the Study of Eco-development of Himalayan Region. Ibid., 132 pp.

PC. 1981a *Report on Development of Backward Hill Areas.* National Committee on Development of Backward Areas. Ibid. (mimeographed), 237 pp.

PC. 1981b. *Report on Development of Tribal Areas.* National Committee on the Development of Backward Areas. Ibid. (cited from Tewari, 1986).

Penafiel, S.R. 1987. Agroforestry research and development in the Philippines. In: P.K. Khosla and D.K. Khurana (eds.), *Agroforestry for Rural Needs,* Vol. I. Proc Intn Workshop, ISTS, pp. 53-61.

Pillai, V.N.K. 1984. Agroforestry relevance and scope. In: R.S. Mathur and M.G. Gogate (eds.), *Agroforestry in India,* op. cit., FRI & C, pp. 76-80.

Pokhriyal, F.C. and G.P. Maithani. 1989. Techniques and approaches to utilize nitrogen fixing systems in forestry plantation programmes. *Indian Forester* 115:136-159.

Pradhan, I.P. 1977. Annual Report, CSWCRTI (mimeographed).

Pradhan, K. 1976. Personal communication (cited from Sen et al., 1978).

Pradhan, K. and P.C. Gupta. 1988. Forage evaluation—Principles and techniques. In: Panjab Singh (ed.), *Pasture and Forage Crops Research,* Range Management Soc. India, IGFRI, pp. 432-440.

Prajapati, M.C. 1974. Report on achievements (1955-74) of Soil Conservation Research, Demonstration and Training Center, Agra, India (mimeographed, data, M.C. Prajapati and S.S. Sajwan).

Prajapati M.C., B. Verma, S.P. Mittal, K.T.N. Nambiar and B.S. Thippannavar. 1971. Effect of lateral develoment of *Prosopis juliflora* D.C. roots on agricultural crops. *Annals Arid Zone* 10 (2&3): 186-193.

Prakash, I. 1981. Important food centre for Indian desert animals, *Bordi (Zizyphus nummularia).* In: H.S. Mann and S.K. Saxena (eds.), Monograph No. 13, CAZRI, pp. 48-50.

Prasad, A. and B. Verma. 1987. Intercropping of field crops with fodder crop of *subabul* (*Leucaena leucocephala*) under rainfed conditions. Annual Report, CSWCRTI, pp. 141-142.

Prasad, R.N., Pati Ram, R.C. Barooah and Munna Ram. 1981. Soil fertility management in north eastern hill region. Res Bull 9, ICAR Res. Comp. NEHR, 30 pp.

Puri, D.N. and P. Joshie. 1979. Economic utilization of class V,VI and VII lands for raising fuel (*Eucalyptus hybrid*) and fodder (*Chrysopogon fulvus*). Annual Report, CSWCRTI.

Puri, S., S. Singh and A. Khera. 1992. Effect of windbreak on the yield of cotton crop in semi-arid regions of Haryana. *Agroforestry Systems* 18: 183-195.

Purohit, M.L. and Wajid Khan. 1981. Socio-economic aspects (of bordi—*Zizyphus nummularia*). In: H.S. Mann and S.K. Saxena (eds.), Monograph No. 13, CAZRI, pp. 59-63.

Purohit, M.L. and Wajid Khan. 1980. Socio-economic dimensions of Khejri. In: H.S. Mann and S.K. Saxena (eds.), *Khejri (Prosopis cineraria) in the Indian Desert*, CAZRI, pp. 56-63.

Purshotham, A. 1962. Utilization of bamboos. *J Timber Dryers and Pres. Assoc. India* 9: 2-19 (cited from Gupta, 1991).

Raizada, M.B. and R.N. Chatterjee. 1956. World distribution of bamboos with special reference to the Indian species and their more important uses, India. Forest Leaflet No 151. Manager of Publications, Govt. of India, Delhi, India (cited from Gupta, 1991).

Ralhan, P.K. 1993. Personal communication.

Ralhan, P.K., A. Singh and R.S. Dhanda. 1993. Performance of wheat as intercrop under poplar (*Populus deltoides* Bertr.) plantation in Punjab, India. *Agroforestry Systems* 19: 217-222.

Ram Parkash. 1983. Personal communication.

Ram Parkash and D. Hocking. 1986. *Some Favourite Trees for Fuel and Fodder.* International Book Distributors, Dehradun, India, 187 pp.

Ramakrishnan, P.S. 1992. *Jhum:* Is there a way out? In: A. Agarwal (ed.), *The Price of Forests,* CSE, pp. 304-310.

Ramakrishnan, P.S., R.P Shukla and Ram Bhoojh. 1982. Growth strategies of trees and their applications to forest management. *Current Science* 51(9): 448-455.

Ramakrishnan, P.S. and O.P. Toky. 1978. Preliminary observations on the impact of shifting agriculture on the forested ecosystem. In: *National Seminar on Resources Development and Environment in the Himalayan Region*, Govt. of India, New Delhi, India, pp. 343-354.

Ramakrishnan, P.S., O.P. Toky, B.K. Mishra and K.G. Saxena. 1978. Slash and burn agriculture in northeastern India. *Proc Fire Regimes & Ecosystem Properties*, East West Center, Honululu, USA, pp. 570-585.

Rambabu, K.G. Tejwani and M.C. Agrawal. 1977. Rainfall erosion index (E.I Value). Annual Report, CSWCRTI, pp. 18-21.

Rana, D.S., G.P. Lodhi, K.D. Taneja, B. Das and D.S. Jatsra. 1988. Effect of spacing on fodder yield and quality of different varieties of *Subabul* and *Anjan* grass under semi-arid conditions. In: Panjab Singh and R.S. Pathak (eds.), *Rangelands—Resources and Management,* op. cit., IGFRI, pp. 333-337.

Randhawa, M.S. 1959. *Farmers of India.* ICAR.

Rao, C.S. 1984. Social forestry in Andhra Pradesh. In: R.S. Mathur and M.G. Gogate (eds.), *Agroforestry in India,* op. cit., FRI & C, pp. 15-21.

Rao, M.R., C.K. Ong, P. Pathak and M.M. Sharma. 1991. Productivity of annual cropping and agroforestry systems on a shallow alfisol in semi-arid India. *Agroforestry Systems* 15: 51-63.

Rawat, G.S. and W.A. Rodgers. 1988. The alpine meadows of Uttar Pradesh: An ecological review. In: Panjab Singh and R.S. Pathak (eds.), *Rangelands—Resources and Management,* op. cit., IGFRI, pp.119-137.

Raychaudhri, S.P., R.R. Agarwal, N.R. Datta Biswas, S.P. Gupta and P.K. Thomas. 1963. *Soils of India.* ICAR.

Reddy, C.V.K. 1981. Agroforestry in coastal Andhra Pradesh. In: *Proc Agroforestry Seminar,* ICAR, pp. 77-82.

Rowbotham, C.J. 1924. The taungya system in Cachar division. *Indian Forester* 50 (7): 356-358.

Rowntree, J.B. 1940. Grazing versus burning as an aid to *Sal* regeneration. *Indian Forester* 66:645-653.

Roy, A.R. 1974. Intercropping in arecanut gardens of north eastern regions of India. A brief review of work done. *Arecanut and Spices Bull* 5 (3): 82-85 (cited from Muralidharan and Nayar, 1979).

Roy, B.B. and A.K. Sen. 1968. Soil map of Rajasthan. *Annals Arid Zone* 7: 1-4.

Sadanandan, A.K. 1974. Raise intercrops in arecanut plantations for higher returns. *Arecanut and Spices Bull* 5 (3): 70-72 (cited from Muralidharan and Nayar, 1979).

Sagwal, S.S. and N.K. Gupta. 1984. Role of aromatic and medicinal plant wealth in forestry extension of Himachal Pradesh. In: R.S. Mathur and M.G. Gogate (eds.), *Agroforestry in India,* op. cit., FRI & C, pp. 474-479.

Saha, N. 1980. Levels of production and income under shifting cultivation in northeast India. In: *Shifting Cultivation in North East India,* NICSSR, pp. 19-27.

Sahasarnaman, K.N. 1964. It pays to grow groundnut in coconut gardens. *Coconut Bull* 18 (4): 123-130 (cited from Gopalasundaram and Nelliat, 1979).

Sahasarnaman, K.N. and N.G. Pillai. 1976. Mixed farming in *coconut garden. Intensive Agric* 4 (9): 6-10.

Sahu, S.B., A. Singh and M.D. Singh. 1980. Farm power sources and their utilization in north eastern region. Res Bull 7. ICAR Res. Comp. NEHR, 22 pp.

Saini, B.C. and K.K. Misra. 1992. With two crops can you also grow trees. *Agroforestry Today* 4(1):17.

Samraj, P. 1977. The blue gum (*Eucalyptus globulus* Labill) tree-An ideal species for farm forestry in the higher hills of the Nilgiris. *Soil conserv Digest* 5(2): 48-52.

Samraj, P., B. Haldorai and C. Henry. 1982. Economic utilization by intercropping in afforested area. Annual Report, CSWCRTI, pp. 47-49.

Sangal, P.M. 1984. Species compatibility considerations in agroforestry: The state of the art in India. In : R.S. Mathur and M.G. Gogate (eds.), *Agroforestry in India,* op. cit., FRI & C, pp. 416-428.

Sannamarappa, M. and A. Muralidharan. 1982. Multiple cropping (in arecanut). In: K.V.A. Bavappa, M.K. Nair and T. Prem Kumar (eds.), Monograph Series No. 2, CPCRI, pp. 33-49.

Sannamarappa, M., S.N. Sampath Kumar and R.K. Singh. 1982. Arecanut, with other crops. Annual Report, CPCRI, pp. 49-51.

Saraswathy, N. and N. Prakasam. 1982. Disease problems in arecanut. *Indian Farmg* 32(9): 29-33, 37.

Sarma, B.S.K. 1984. Agrogorestry in Manipur. In: R.S. Mathur and M.G. Gogate (eds.), *Agroforestry in India,* op. cit., FRI & C, pp. 138-146.

Sathees Babu, K., D. Jose and C. Gokulpalan. 1992. Species diversity — In a Kerala home garden. *Agroforestry Today* 4(3): 15.

Saxena, K.G. and P.S. Ramakrishnan. 1982. Partitioning of biomass and nutrients in the secondary successional herbaceous population subsequent to slash and burn. *Proc Indian Nat. Sci. Acad.* B48(6): 807-818.

Saxena, N.C. 1991. Marketing constraints of *Eucalyptus* from farm lands in India. *Agroforestry Systems* 13: 73-85.

Saxena, S.K. 1984 *Khejri* (*Prosopis cineraria*) and *bordi* (*Zizyphus nummluaria,* multipurpose plants of arid and semi-arid zones. In: K.A. Shankarnarayan (ed-in-chief), *Agroforestry in Arid and Semi-arid Lands,* op. cit., CAZRI, pp. 111-121.

Saxena, S.K. 1981. *Morphology and ecology of bordi* (*Zizyphus nummularia*). In: H.S. Mann and S.K. Saxena (eds.), *Bordi (Zizyphus nummularia),* Monograph No. 3, CAZRI, pp. 3-9.

Saxena, S.K. 1980. Taxonomy, morphology, growth and reproduction of *Khejri* and its succession in northwest India. In: H.S. Mann and S.K. Saxena (eds.), *Khejri (Prosopis cineraria) in the Indian Desert,* CAZRI, pp. 4-10.

Saxena, S.K. 1977. Vegetation and its succession in the Indian desert. *Desertification and Its Control,* ICAR, pp. 176-192.

Saxena, S.K. and S.K. Sharma. 1981. Productivity of the shrub. In: H.S. Mann, and S.K. Saxena (eds.), *Bordi* (*Zizyphus nummularia*), op. cit., CAZRI, pp. 25-30.

Sen, K.C. 1953. *Animal Nutrition Research in India*. Macmillan & Co. Ltd., Calcutta, India, 370 pp.

Sen, K.C., S.N. Ray and K. Ranjhan. 1978. *Nutritive Values of Indian Cattle Feeds and the Feeding of Indian Animals*. ICAR, New Delhi, 92 pp.

Seshachalam, N. 1979. Personal communication.

Seth, S.K. 1981. *India and Sri Lanka: Agroforestry for Local Community Development Programme*. FAO, Rome, GCP/INT/347/SWE, 116 pp.

Shah, M.H. 1988. Role of nomads in the destruction of alpine and subalpine pastures and future strategies. In: Panjab Singh and R.S. Pathak (eds.), *Rangelands—Resources and Management*, op. cit., IGFRI, pp.148-152.

Shah, S.L. 1981. Agricultural planning and development in the northwestern Himalayas, India. In: *Nepal's Experience in Hill Agricultural Development*, Ministry Food Agri., Kathmandu, Nepal, pp. 160-168.

Shankar V. 1984. Interrelationship of tree overstorey and understorey vegetation in silvipastoral system. In: K.A. Shankaranayan (ed.-in-chief), *Agroforestry in Arid and Semi-arid Lands*, CAZRI, pp. 143-149.

Shankar, V. 1981. *Zizyphus nummularia* in the Indian scriptures. In: H.S. Mann and S.K. Saxena (eds.), *Bordi* (*Zizyphus nummularia*), op. cit., CAZRI, pp. 1-2.

Shankar, V. 1980a. *Khejri* in Indian scriptures. In: H.S. Mann and S.K. Saxena (eds.), *Khejri (Prosopis cineraria) in the Indian Desert*, op. cit., CAZRI, pp. 1-3.

Shankar, V. 1980b. Distribution of *Khejri* (*Prosopis cineraria Macbride*) in western Rajasthan. In: ibid., pp. 11-19.

Shankar, V., Suresh Kumar and R.K. Tyagi. 1988. Grazing resources of semi-arid and arid regions. In: Panjab Singh (ed.), *Pasture and Forage Crops Research*, op. cit., IGFRI, pp.63-85.

Shankarnarayan,K.A. (ed.-in-chief). 1984. Agroforestry in Arid and Semi-arid Zones. Pub No. 24, CAZRI, 295 pp.

Shankarnarayan, K.A., L.N. Harsh and S. Kathju. 1987. Agroforestry in the arid zones of India. *Agroforestry Systems* 5: 69-88.

Shanmugasundaram, S. 1984. Agroforestry—its aspects and prospects in Tamilnadu. In: R.S. Mathur and M.G. Gogate (eds.), *Agroforestry in India*, op. cit., FRI & C, pp. 186-192.

Sharma, B.K. and R.P. Dhir. 1981. Distribution of trace elements in foliage. In: H.S. Mann and S.K. Saxena (eds.), *Bordi* (*Zizyphus nummularia*), CAZRI, pp. 34-37.

Sharma, D.D. and R.S. Gill 1969. Chemical composition and nutritive value of *biul* (*Grewia oppositifolia* Roxb) tree leaves. *J Res Punjab Agric Univ* 3(2): 178-185.

Sharma, K. and N.K. Bhattarcharyya. 1988. Integrating agroforestry with goat husbandry for the benefit of the rural poor. *Wastelands News* 4(1) :13-16.

Sharma, K., N. Singh, J.L. Ogra and N.K. Bhattacharyya. 1988. Goat production from grazing and browsing resources. In: Panjab Singh (ed.), *Pasture and Forage Crops Research*, op. cit., IGFRI, pp. 423-431.

Sharma, K.K. 1992. Wheat cultivation in association with *Acacia nilotica*(L) Wild ex Del. Field bund plantation—a case study. *Agroforestry Systems* 17: 43-51.

Sharma, T.C. 1980. The prehistoric background of shifting cultivation. In: *Shifting cultivation in Northeast India*, NICSSR, pp. 1-5.

Shaukat Hussain. 1933. Raising *sal* by departmental taungya in UP. *Indian Forester* 59 (5): 320-321.

Shaukat Hussain. 1925a. The development of *sal* seedlings in Gorakhpur taungyas. *Indian Forester* 51 (2): 69-72.

Shaukat Hussain. 1925b. Taungya in Gorakhpur forest division, Eastern Circle, UP. *Indian Forester* 51 (3): 114-116.

Shaukat Hussain. 1925c. Suitable field crops for *sal* taungya in UP. *Indian Forester* 51(5): 192-198.

Sheshadri, P. 1976. *Acacia leucophloea. Forestry Scientific Workers Conference*. Tamil Nadu Agricultural University, Coimbatore, India (cited from Jambulingam and Fernandes, 1986).

Singh, A. 1981. Shifting cultivation. In: A.N. Asthana, S.P. Ghosh, P.S.R.C. Murti and R.N. Verma (eds.), *Proc Workshop Agric Res in Northeastern Hills Region*, ICAR Res. Com. NEHR, pp. 174-178.

Singh, A. and M.D. Singh. 1981. Soil erosion hazards in northeastern hill region. Bull 10, ICAR Res. Com. NEHR, 30 pp.

Singh, A. and M.D. Singh. 1980. Effect of various stages of shifting cultivation on soil erosion from steep slopes. *Indian Forester* 106: 106-121.

Singh, B. 1971. A comparative study on economics of various soil conservation cum grassland improvement practices for rejuvinating forage production in ravine lands, I (forage production). *Indian Forester* 97: 315-321.

Singh, G.B. and P.O. Pazo. 1981. Agroforestry in eastern Himalaya. *Proc Agroforestry Seminar*, ICAR, pp. 27-39.

Singh, H. and P. Joshie. 1984. Economic utilization of bouldery riverbed land for fodder, fibre and fuel under rainfed conditions. Annual Report, CSWCRTI, pp. 44-46.

Singh H., M.K. Vishwanatham, P. Joshie and S.P. Bhardwaj. 1990a. Effect of spacing and grass combination on fodder yield of *subabul* in bouldery class V-VII land under rainfed conditions. Annual Report, 1989-90, ibid., pp. 45-46.

Singh H., M.K. Vishwanatham, P. Joshie and S.P. Bhardwaj. 1988a. Effect of spacing and grass combination on fodder yield of *subabul* in bouldery class V-VII land under rainfed conditions. Annual Report, ibid., pp. 37-38.

Singh H., M.K. Vishwanatham, P. Joshie and S.P. Bhardwaj. 1987a. Effect of spacing and grass combination on fodder yield of *subabul* in bouldery class V-VII land under rainfed conditions. Annual Report, ibid., pp. 49-50.

Singh H., M.K. Vishwanatham, Joshie, P. and S.P. Bhardwaj. 1986b. Effect of spacing and grass combination on fodder yield of *subabul* in bouldery class V-VII land under rainfed conditions. Annual Report, ibid., pp. 45-46.

Singh, H., M.K. Vishwanatham, P. Joshie and M.L. Khybri. 1991. Economic utilization of bouldery riverbed land for fodder, fibre and fuel. Annual Report, 1990-91, ibid., pp. 23-24.

Singh, H., M.K. Vishwanatham, P. Joshie and M.L. Khybri. 1990b. Economic utilization of bouldery riverbed land for fodder, fibre and fuel. Annual Report, 1989-90, ibid., pp. 41-42.

Singh, H., M.K. Vishwanatham, P. Joshie and M.L. Khybri. 1988b. Economic utilization of bouldery riverbed land for fodder, fibre and fuel. Annual Report, ibid., pp. 36-37.

Singh, H., M.K. Vishwanatham, P. Joshie and M.L. Khybri. 1987b. Economic utilization of bouldery riverbed land for fodder, fibre and fuel. Annual Report, ibid., pp. 48-49.

Singh M.C., A. Singh, and S.D. Sahu. 1980. Traditional farm tools of northeastern hill region. Res Bull 5, ICAR, Res. Comp. NEHR, 58 pp.

Singh, P. 1988. Rangeland reconstruction and management for optimizing biomass production. In: Panjab Singh and P.S. Pathak (eds.), *Rengelands—Resources and Management*, op. cit., IGFRI, pp. 20-37.

Singh, R., N. Yadkumar, K.N. Roy Burman and A.C. Roy. 1982. Intercropping in arecanut gardens in north Bengal. *Indian Farmg* 32 (9): 13-15.

Singh, R.K. and A.K. Roy. 1983. Economics of inter and mixed cropping in arecanut gardens in research centers and farmers fields, in West Bengal and Assam. Annual Report, CPCRI, pp. 57-58.

Singh, R.P., K. Vijayalakshmi, G.R. Korwar and Mohd. Osman. 1987c. Alternate land use systems for drylands of India. Res Bull 6, Central Res. Inst. Dryland Agri., Hyderabad, India, 61 pp.

Singh, R.V. 1991. Small scale multipurpose production systems in a mountainous region of India. Field Doc. No. 28. Regional Wood Energy Dev. Prog. Asia, FAO Regional Office, Bangkok, Thailand, 67 pp.

Singh, S., S. Krishnamurthi and S.L. Katyal. 1963. *Fruit Cultivation in India*. ICAR, 451 pp.

Singh, S.P., Jeetram and R.P. Singh. 1988c. Forage production and utilization in Central Himalayan rangelands. In: Panjab Singh (ed.), *Pasture and Forage Crops Research*, op. cit., IGFRI, pp. 86-96.

Singh, V., Pandey, P.C., Shri Pal and Mahendra Pal. 1986a. Studies on the yield and quality of temperate forage grasses and legumes and their effect on apple trees and soil nitrogen under rainfed conditions of Kumaun hills of UP, India. *Himalayan Res & Development* 5(I & II): 10-13.

Sinha, R.L. 1958. Fodder resources of U.P. *Southern Forest Rangers College Magazine.* 34: 180-190 , cited from M. Singh. (ed.) 1983. Top Feed Resources—Their Production, Utilization and Conservation, Central Sheep Wool Res. Inst., Avikanagar, India, pp. 1-13.

Sowani, M.Y. 1980. Cash crops and taungya cultivation. Second Forestry Conference, FRI & C (cited from Sangal, 1984).

SPB. 1985. Report of the High Level Committee on Land and Water Resources. State Planning Board, Government of Kerala, Trivandrum, India (cited from Nair and Sreedharan, 1986).

Srimathi, R.A. and R.K. Venkatesan. 1981. Agroforestry: Plant resources of the Deccan region. *Proc Agroforestry Seminar,* ICAR, pp. 210-219.

Srivastava, J.B.L. 1984. Practice of agroforestry with special reference to Haryana. In: R.S. Mathur and M.G. Gogate (eds.), *Agroforestry in India,* op. cit., FRI & C, pp. 81-84.

Srivastava J.P.L. 1978. Lopping studies on *Prosopis cineraria. Indian Forester* 104(4): 269-274.

Srivastava, M.B., M. Srivastava and C.B. Lal. 1988. Forage and livestock development in middle Himalaya. In: Panjab Singh and R.S. Pathak (eds.), *Rangelands—Resources and Management,* op. cit., IGFRI, pp.153-161.

Srivastava, M.D. 1984b. Economics of *Eucalyptus* planting in agroforestry in Haryana with special reference to Hissar. In: R.S. Mathur and M.G. Gogate (eds.), *Agroforestry in India,* op. cit., FRI & C, pp. 347-351.

Stewart, D. 1933. Financial prospects of departmental taungyas in Haldwani, UP. *Indian Forester* 59 (3):142-148.

Subbayan, R. 1984. Studies on management of *Prosopis juliflora* for fuel and fodder. Annual Report, CSWCRTI, pp. 159-160.

Subbayan, R. and S. Chittaranjan. 1979. Fodder-cum-fuel plantation in the black cotton soils of Bellary under rainfed conditions. Annual Report, ibid., pp. 57-58.

Suchiang, F. 1984. Agroforestry in Meghalaya. In: R.S. Mathur and M.G. Gogate (eds.), *Agroforestry in India,* op. cit, FRI & C, pp. 147-153.

Sud, A.D. and S.P. Mittal. 1986. Tie ridge technique of raising *Eucalyptus* and *Bhabar* on 4% slope. Annual Report, CSWCRTI, pp. 83-84.

Talbot, W.A. 1909 and 1912. *Forest Flora of Bombay Presidency and Sindh.* Vol. I (1909) and Vol. II (1912). Reprinted 1976. Today & Tomorrow Printers and Publishers, New Delhi, India, 508 pp. (Vol. I) and 574 pp. (Vol II).

Tejwani, K.G. 1989. Agroforestry practices in hill regions of India. In : T.V.Singh and Jagdish Kaur (eds.), *Studies in Himalayan Ecology,* Himalayan Books, New Delhi, India, pp. 201-214.

Tejwani, K.G. 1987a. Agroforestry practices and research in India. In: H.L. Gholz (ed.), *Agroforestry: Realities, Possibilities and Potentials,* Martinus Nijhoff Publisher, Dordrecht, Netherlands, pp. 109-136.

Tejwani, K.G. 1987b. Agroforestry in hill regions of India. Proc International Workshop Agroforestry for Rural Needs. In: D.K. Khurana and P.K. Khosla (eds.), *Agroforestry for Rural Needs,* Vol.II, 1993, ISTS, pp. 389-407.

Tejwani, K.G. 1984a. Biophysical and socio-economic causes of land degradation and a strategy to foster watershed rehabilitation in the Himalayas. In: C.L.O. Loughlin and A.J. Pearce (eds.), *Symposium on Effects of Forest Land Use on Erosion and Slope Stability,* East West Center, Honolulu, USA, pp. 55-59.

Tejwani, K.G. 1984b. Agroforestry in Asia Pacific. Working paper Environment and Policy Institute, East West Center, Honolulu, USA (mimeographed) 111 pp.

Tejwani, K.G. 1981. Agroforestry for the wasted lands in western Himalayas In: *Proc Agroforestry Seminar,* ICAR, pp. 220-233.

Tejwani, K.G. and Chun, K. Lai. 1992. Asia-Pacific Agroforestry Profiles. FAO Asia-Pacific Agroforestry Network, Field Doc. No 1, Bogor, Indonesia, 140 pp.

Tejwani, K.G., V.V. Dhruvanarayana and T. Satyanarayana. 1960. Control of gully erosion in ravine lands of Gujarat. *J Soil Water Conserv India* 8: 65-74.

Tejwani, K.G., V. Srinivasan and M.S. Mistry. 1961. Gujarat can save its ravine land. *Indian Farmg* 11: (8) 20-21.

Tewari, D.N. 1986. *Forest and National Development.* Jugal Kishore & Co., Dehradun, India, 256 pp.

Toky, O.P. and R.P. Bisht. 1992. Observations on rooting patterns of some agroforestry trees in arid regions of northwestern India. *Agroforestry Systems* 18: 245-264.

Toky, O.P. and P.S. Ramakrishnan. 1982. A comparative study of the energy budget of hill agro-ecosystems with emphasis on slash and burn system (*Jhum*) at lower elevations of northeastern India. *Agroecosystem* 9(2): 143-154.

Toky, O.P. and P.S. Ramakrishnan. 1981. Cropping and yields in agricultural systems of the northeastern hill regions of India. *Agroecosystem* 7: 11-25.

Tomar, G.S. and S.K. Shrivastava. 1986. Preliminary studies of rice cultivation in association with trees. In: P.K. Khosla., Sunil Puri and D.K. Khurana (eds.), *Agroforestry Systems—A New Challenge,* Proc Satelite Seminar on Agroforestry Systems—A New Challenge, ISTS, pp. 207-211.

Trafford, F. 1924. Sugar cane in taungyas. *Indian Forester* 50 (9): 494-495.

Tripathi, K.P. 1984. Agroforestry in north Bihar. In: R.S. Mathur and M.G. Gogate (eds.), *Agroforestry in India,* op. cit., FRI & C, pp. 27-34.

Troup, R.S. 1921. *The Silviculture of Indian Trees.* Vol. I, II and III, Clarendon Press, London, England.

Tyagi, R.S. and Panjab Singh. 1988. Grazing resources and grazing systems in India. In: Panjab Singh (ed.), *Pasture and Forage Crops Research,* op. cit., IGFRI, pp.17-34.

Unadkat, V. 1984. Agroforestry in Gujarat. In: R.S. Mathur and M.G. Gogate (eds.), *Agroforestry in India,* op. cit., FRI & C, pp. 35-57.

Upadhaya, R.C. and S.P. Ghosh. 1983. Wild cardamom of Arunachal Pradesh. *Indian Hort* 27 (4): 25-27.

Van Soest, P.J. 1967. Development of comprehensive system of feed analysis and its application to forages. *J Animal Sci* 26: 119-127 (cited from Pradhan and Gupta, 1988).

Varghese, P.T., P.K.R. Nair, E.V. Nelliat, Rama Verma and P. Gopalasundaram. 1978a. Intercropping with tuber crops in coconut gardens. *Proc Placrosym I,* Placrosym Standing Committee, Kasargod, India, pp. 399-405.

Varghese, P.T., E.V. Nelliat and T.K. Balakrishnan. 1978b. Beneficial interactions of coconut cacao crop combination. Ibid., pp. 383-392.

Varma, A., B.P.S. Yadav, K.T. Sampath and D.J. Roy. 1983. *J Res Assam Agric Univ* 4(2): 117-124.

Varma, A., B.P.S. Yadav, K.T. Sampath and D.J. Roy. 1982a. Livestock Feed and Feeding Habit in Northeastern Hills of India. Part I, Indigenous feeds and resultant productivity, Res Bull 17, ICAR Res. Comp. NEHR, 100 pp.

Varma, A., B.P.S. Yadav, K.T. Sampath and D.J. Roy. 1982b. Chemical composition of fodder tree leaves, shrubs and epiphytes in Northeastern Hill Region, *Indian J Animal Sci* 52 (10): 859-866.

Vellapan, E. and C.K. George. 1982. Development (of arecanut). In: K.V.A. Bavappa., M.K. Nair and T. Prem Kumar. Monograph series No. 2, CPCRI, pp. 305-318.

Venkataramani, S.K. 1973. Shade for tea in relation to environment. *Planters Chron* 68 (13): 267-274.

Venugopal, K. 1986. Prospects of agroforestry in Sikkim. In: P.K. Khosla., Sunil Puri and D.K. Khurana (eds.), *Agroforestry Systems—A New Challenge.* op. cit., ISTS, pp. 69-74.

Vergara, N.T. (ed.). 1982. *New Directions in Agroforestry: The Potential of Tropical Legume Trees.* East West Center, Honolulu, USA, 55pp.

Watson, H.R. and W. Laquihan. 1985. Sloping agricultural land technology (SALT) as developed by Mindanao Baptist Rural Life Center. Workshop on Site Protection and Amelioration Roles of Agroforestry. University of Philippines, Los Banos, Philippines (cited from Penafiel, 1987).

Watt, G. 1893. *A Dictionary of the Economic Products of India.* Delhi, India (cited from K.S. Bhat, 1979).

Watt. G. and H.H. Mann. 1903. *The Pests and Blights of the Tea Plant.* 2nd edition. Govt. of India, Calcutta, India (cited from Hadfield, 1974).

de Weille, G.A. 1959. Adaptation of tea cultivation to the occurrence of blister blight. *Arch Tea Culti* 20: 161-192 (cited from Venkataramani, 1973).

Wickramsinghe, A. 1991. Gender issues in the management of home gardens: A case study of Kandyan home garden in Sri Lanka. *International Symposium on Man-made Community Integrated Land Use and Biodiversity in the Tropics,* Academia Sinica, Xishuanbanna, China.

Wiersum, K.F. 1986. Social forestry and agroforestry in India. Report No. 456, Dept Forest Management, Agric. University, Wegeningen, Netherlands (mimeographed), 33 pp.

Wiersum, K.F. 1982. Tree garden and taungya on Java. Examples of agroforestry in the humid tropics. *Agroforestry Systems* 1: 53-67.

Wight, W. 1959a. The shade tree tradition in the tea gardens of north India: I. the value of shade. Annual Report, Indian Tea Assoc. Sci. Dept. Tocklai Expt. Sta., Jorhat, India, pp. 75-98.

Wight, W. 1959b. The shade tree tradition in the tea gardens of north India: II. The use of ammonium sulphate. Annual Report, ibid., pp. 99-122.

Wight, W. 1959c. The shade tree tradition in the tea gardens of north India III. Superphosphate and light intensity. Annual Report, ibid., pp. 123-150.

Wood, B.R. 1924. Field crops in taungyas. *Indian Forester* 50 (7): 391-395.

Wood, G.A.R. 1964. *Cacao Growing in India.* Cadbury Brothers Ltd. Bournville, England (cited from Bhat and Bavappa, 1972).

SUBJECT INDEX

AUTHOR INDEX

SPECIES INDEX